NUMERICAL COGNITION AND THE EPISTEMOLOGY OF ARITHMETIC

Arithmetic is one of the foundations of our educational systems, but what exactly is it? Numbers are everywhere in our modern societies, but what is our knowledge of numbers really about? This book provides a philosophical account of arithmetical knowledge that is based on the state-of-the-art empirical studies of numerical cognition. It explains how humans have developed arithmetic from humble origins to its modern status as an almost universally possessed knowledge and skill. Central to the account is the realisation that, while arithmetic is a human creation, the development of arithmetic is constrained by our evolutionarily developed cognitive architecture. Arithmetic is a sophisticated cultural development, but it is ultimately based on abilities with numerosities that we already possess as infants and share with many non-human animals. Therefore, arithmetic is not purely conventional, an arbitrary game akin to chess. Instead, arithmetic is deeply connected to our basic cognitive capacities.

MARKUS PANTSAR is a visiting professor at the RWTH Aachen University, Germany, and docent in theoretical philosophy at the University of Helsinki, Finland. He is the author of *Truth, Proof, and Gödelian Arguments: A Defence of Tarskian Truth in Mathematics* (2009) and many journal articles on the cognitive foundations of arithmetical knowledge.

NUMERICAL COGNITION AND THE EPISTEMOLOGY OF ARITHMETIC

MARKUS PANTSAR
RWTH Aachen University

Shaftesbury Road, Cambridge CB2 8EA, United Kingdom
One Liberty Plaza, 20th Floor, New York, NY 10006, USA
477 Williamstown Road, Port Melbourne, VIC 3207, Australia
314–321, 3rd Floor, Plot 3, Splendor Forum, Jasola District Centre, New Delhi – 110025, India
103 Penang Road, #05-06/07, Visioncrest Commercial, Singapore 238467

Cambridge University Press is part of Cambridge University Press & Assessment, a department of the University of Cambridge.

We share the University's mission to contribute to society through the pursuit of education, learning and research at the highest international levels of excellence.

www.cambridge.org
Information on this title: www.cambridge.org/9781009468886

DOI: 10.1017/9781009468862

First published 2024

A catalogue record for this publication is available from the British Library

A Cataloging-in-Publication data record for this book is available from the Library of Congress

ISBN 978-1-009-46888-6 Hardback

Contents

Figures

Preface

This book has been long in the making, in fact for many years before I explicitly decided to write a book about numerical cognition and the epistemology of arithmetic. Its genesis can be traced back to reading three books during my master's studies: the first edition of Stanislas Dehaene's (1997) *The Number Sense*, Brian Butterworth's (1999) *What Counts* (published as *The Mathematical Brain* in the UK) and *Where Mathematics Comes From* by George Lakoff and Rafael Núñez (2000). I had completed my PhD in 2009 with a thesis on Tarskian truth and Gödel's incompleteness theorems (Pantsar, 2009). This logic-based approach to the foundations of mathematics was – and still is – fascinating to me, but in my first postdoctoral project I started to feel like I was running out of interesting things to say about it. Meanwhile, the three books I had read about the cognitive foundations of mathematics started to haunt me. Initially I had read them as an interested outsider – after all, my background was in philosophy and mathematics, not in the cognitive sciences – but now I read them with the mindset of exploring the philosophical possibilities of empirical approaches to mathematical cognition. Quickly I realised that the potential was enormous.

Switching my focus to the cognitive side of mathematics was a curious experience. I received a lot of sceptical feedback concerning the philosophical relevance of the cognitive origins of mathematics. I could understand that very well. When reading the work of Dehaene, Butterworth and others, I was constantly troubled by the cavalier attitude they showed towards the kind of conceptual distinctions that I held sacrosanct as a philosopher educated in the analytic tradition. Passages about infants knowing arithmetical truths, for example, were frustrating to read. Nevertheless, I became increasingly convinced that these authors were on to something. Mathematics *is* based in an important way on evolutionarily developed abilities, and these abilities need to be studied empirically.

This was crucial to acknowledge in order to make progress in the epistemology of mathematics. The challenge was then to make sense of the empirical data with systematic conceptual distinctions in place. This book is the culmination of more than a decade of work in response to that challenge in the field of arithmetic.

By the time I published my first papers on the topic, the landscape was already changing. Indeed, Marcus Giaquinto – one of the pioneers in the empirically informed philosophy of mathematics – once commented after a talk I gave that I was pushing at an open door: philosophers had already by and large accepted that empirical data on numerical cognition is philosophically relevant. Fortunately, he was correct, and the positive trend has continued. I have noticed that the general attitude among philosophers of mathematics has become increasingly accepting of empirically informed approaches. More philosophers work on such approaches than ever before, and there have been important developments in the field in the past ten years. Indeed, there have also been important developments in my own work. Admittedly, I am not happy with some of the things I wrote in my first papers. I focused too much on the evolutionarily developed cognitive origins of arithmetic and not enough on the importance of cultural influences. It was only during the last few years that I started to form a comprehensive idea of how the evolutionarily developed quantitative abilities develop into proper arithmetical abilities. And it was only after getting this comprehensive idea that I could properly assess its importance for the epistemology of arithmetic.

While my earlier work leaves room for improvement, it was sufficiently on the right track to be used as a partial basis for this book project. While most of the content is new, there are parts where it made sense to apply some of my previously published work. Chapter 1 draws from 'An empirically feasible approach to the epistemology of arithmetic' (Pantsar, 2014) and 'In search of aleph-null: How infinity can be created' (Pantsar, 2015b). Chapter 2 draws partly from 'Bootstrapping of integer concepts: The stronger deviant-interpretation challenge (and how to solve it)' (Pantsar, 2021a), and Chapter 3 from 'The enculturated move from proto-arithmetic to arithmetic' (Pantsar, 2019b) and 'A fresh look at research strategies in computational cognitive science: The case of enculturated mathematical problem solving' (Fabry and Pantsar, 2021). 'Early numerical cognition and mathematical processes' (Pantsar, 2018a) forms some of the basis for Chapter 5. Chapter 8 is partly based on 'Objectivity in mathematics, without mathematical objects' (Pantsar, 2021c),

'The modal status of contextually a priori arithmetical truths' (Pantsar, 2016b) and 'From maximal intersubjectivity to objectivity: An argument from the development of arithmetical cognition' (Pantsar, 2023b). Finally, Chapter 9 is partly based on 'On what ground do thin objects exist? In search of the cognitive foundations of number concepts' (Pantsar, 2023d).

Many people have helped me with this project. First and foremost, I am very grateful to my partner and collaborator Regina Fabry who, in addition to all her other help and support over the years, provided extremely important comments on the manuscript. Bahram Assadian and Stefan Buijsman also provided many comments that were crucial for sharpening the argumentation. In addition to them, Mirja Hartimo gave me important advice. I acknowledge all their help with great gratitude, as I do the many extremely helpful comments received from the two anonymous reviewers of the manuscript. I also want to thank everybody who has provided helpful feedback on my earlier papers and presentations on the topic. This includes Marianna Antonutti, Sorin Bangu, Fausto Barbero, Neil Barton, Daniel Cohnitz, Sorin Costreie, Mircea Dumitru, Catarina Dutilh Novaes, Benedict Eastaugh, José Ferreirós, Marcus Giaquinto, Valeria Giardino, Alex Gillett, Severi Hämäri, Ole Hjortland, Daniel Hutto, Max Jones, Jeffrey Ketland, Brendan Larvor, Hannes Leitgeb, Øystein Linnebo, Benedikt Löwe, Penelope Maddy, Mathieu Marion, Richard Menary, Alin Olteanu, Peter Pagin, Tuomas Pantsar, Jean-Charles Pelland, Paula Quinon, Panu Raatikainen, Erich Reck, Colin Rittberg, Marcus Rossberg, Jenni Rytilä, César dos Santos, Barbara Sarnecka, Glenda Satne, Dirk Schlimm, Andrea Sereni, Frederik Stjernfelt, Tuomas Tahko, Tuukka Tanninen, Fenner Tanswell and Sean Walsh. If I have forgotten anyone, I sincerely apologise. Of course, all errors and inaccuracies in the book are solely my responsibility.

I would also like to give my special thanks to Gabriel Sandu for supporting my choice to move my research focus to the cognitive side of mathematics during my postdoctoral years. This book manuscript was developed during my time as a grant researcher at the University of Helsinki, made possible by a research grant from the Finnish Cultural Foundation, and a period as a senior research fellow at the Käte Hamburger Kolleg 'Cultures of Research', RWTH Aachen University. The manuscript was finalised while working as a guest professor at RWTH Aachen University at the Chair for Theory of Science and

Technology. I want to thank Gabriele Gramelsberger, the holder of the chair, for her support.

Finally, I would like to thank my parents Merja and Lasse for making it possible for me to pursue a career as a researcher in the first place.

Introduction
A Fractured Landscape

There is no overestimating the importance of arithmetic for the development of human cultures. For the majority of people in the modern world, numbers and arithmetic are a constant part of everyday life. From keeping track of time to financial transactions, most modern societies could not function without arithmetic. For that reason, it has also become a crucial part of modern education. Alongside reading and writing, arithmetic is among the first subjects that children are taught as part of their formal education in most cultures.

Yet this was not always so. As we will see, the origins of arithmetic are not easy to trace but, in the scale of human development, it was not that long ago that all cultures were non-arithmetical. Indeed, as we will also see, there are still cultures that are non-arithmetical. This prompts the fundamental question in the history of arithmetic: how did arithmetic develop? If it is a purely cultural development, how could it develop in such similar ways independently in several cultures? If it is not a cultural development, why are there cultures that not only do not practice arithmetic but do not even have numeral words?

The answer I will argue for in this book is that arithmetic is a cultural development, but it is in important ways based on cognitive abilities that are shared universally by humans (as well as many non-human animals). This makes arithmetic philosophically highly interesting, especially in terms of epistemology. Since antiquity, philosophical accounts of arithmetic (and mathematics more generally) were for millennia dominated by views according to which the subject matter of arithmetic is eternal and independent of human cultures. In this book, I develop a radically different account, according to which the subject matter of arithmetic is a cultural creation. This does not mean that arithmetic consists merely of arbitrary conventions. Nor does it mean that arithmetic is not based on culture-independent factors. But it does imply that we need a non-traditional account of the epistemology of arithmetic; arithmetical knowledge can no longer be assumed to concern eternal truths about objectively existing numbers.

From that background, I want to provide an answer to what I see as the most fundamental question in the epistemology of arithmetic: How can arithmetical knowledge be *acquired*? Traditionally, philosophers have focused on the various aspects of the question of what arithmetical knowledge is *like*. These aspects (e.g., apriority, necessity, objectivity) will be important topics for this book. However, I believe that they can be understood better when we have a clearer idea of the way human agents are able to acquire arithmetical knowledge in the first place. Indeed, I believe that traditional epistemology of arithmetic has been hindered by too much focus on analysing the nature of arithmetical knowledge and too little on determining its origins. One main aim of this book is to alter that balance. In doing that, we need to be open to broadening the methodology of epistemology. As I will show, the acquisition of arithmetical knowledge is far from being a question only for philosophy, and proper methodology needs to be sensitive to that.

The question of how arithmetical knowledge is acquired consists of two parts, one concerning the *ontogeny* of arithmetical knowledge and one concerning its *phylogeny* and *cultural history*. The former question asks how human individuals can acquire arithmetical knowledge, while the latter question concerns how human *cultures* can develop arithmetical knowledge. As we will see, these two questions are tightly connected, and a satisfactory epistemological account of arithmetic will need to address both. I will also show that the question of arithmetical knowledge cannot be separated from studying arithmetic as a human endeavour and not simply as knowledge of formal arithmetic, as is often done in philosophy. In addition to knowledge, we need to consider arithmetical practices, skills, tools and applications. Furthermore, all these aspects need to be considered from the perspective of an individual learning arithmetic, as well as the perspective of a culture developing arithmetic.

The division into the ontogenetic and the phylogenetic/historical questions is particularly important because, as I will show in this book, arithmetic as we know it is ultimately a human cultural phenomenon. That matters for two reasons. First, it means that arithmetic should be studied as a cultural phenomenon. There is a great amount of literature about arithmetic in non-human animals and human infants, but I will show that none of that actually concerns arithmetic. While non-human animals and human infants have capacities with numerosities, these should be distinguished from proper arithmetic. Proper arithmetic is a cultural phenomenon and as such it exists only within arithmetical cultures. Studying arithmetic as a cultural phenomenon requires recognising our evolutionarily developed capacities with numerosities – those that we possess already as infants and share with many non-human animals – but it also requires much more. To understand this we need to have a better grasp of how humans develop and acquire arithmetical knowledge. This is what I will focus on in Part I (Ontogeny) and Part II (Phylogeny and History) of this book.

The second reason why it is important to recognise arithmetic to be a human cultural phenomenon is that we do not need anything *further* to understand the epistemology of arithmetic. As I will argue in Part III, all the relevant epistemological questions concerning arithmetic can be answered based on the ontogeny, phylogeny and cultural history of arithmetical knowledge. In addition, I will show that this epistemological analysis is also relevant for the ontology of arithmetic. In this, Part III of the book may be seen as controversial. Traditionally, both the ontology and the epistemology of mathematics, including that of arithmetic, have been considered to be subjects for a priori study. In contrast, my treatment of arithmetical knowledge in Parts I and II is based on empirical findings. However, I do not believe that there is any reason for controversy. In addition to showing what arithmetical knowledge is like as a culturally developed phenomenon, the other main purpose of this book – a *meta*-purpose, if you will – is to show that traditional a priori methodology can fruitfully co-exist with empirically-informed epistemology. Indeed, as Part III seeks to show, we need a synthesis of both in order to understand the epistemology of arithmetic.

Nevertheless, I do recognise that in this book I navigate a fractured landscape. I build on the work of researchers whose views may be inconsistent with each other, and some of them might not be happy with my aims. In the research on arithmetical knowledge, there are different cultures that have different methodologies, different forms of argumentation and different terminology. Taken at face value, the empirical study of numerical cognition and traditional a priori epistemology of arithmetic may seem to be irreversibly incompatible fields of study. Yet that is exactly the kind of attitude that I want to challenge with this book. What I want to show is that the differences in the disciplines are ultimately a richness. But it is a richness we can only recognise if we possess ways of seeing through the superficial differences and find common ground in diverse research paradigms.

In the rest of this introductory chapter, I want to detail how that can be achieved. The main aim of the chapter is to locate my work in the multifaceted literature generally on mathematical and more specifically on arithmetical knowledge, including both empirical and philosophical approaches. In Section I.4, after there has been a chance to describe the methods and concepts that I will be applying, I will present the structure of the book in detail. But, to give an idea of the way things proceed, after the introduction the work is divided into three parts. In Part I, the focus is on the acquisition of arithmetical knowledge in individual ontogeny. In Part II, I move to the question of how arithmetical knowledge has developed in phylogeny and cultural history. In these two parts, I develop an epistemological account of arithmetic that I then assess philosophically in Part III. According to this account, arithmetical knowledge is a cultural development that is made possible by evolutionarily developed abilities for observing our environment in terms of numerosities. As I will argue, however, this does not make the epistemological status of arithmetical knowledge somehow 'weaker'. Instead, the key characteristics traditionally associated with arithmetic – namely: apriority, necessity, objectivity and universality – are included in a sufficiently strong sense in the present account.

Before we start, one final thing should be noted. This book is limited to arithmetic, but I am confident that similar treatments could be given to other fields of mathematics. Indeed, early such work exists already for the development of geometrical cognition (Hohol, 2019; Pantsar, 2021d). Arithmetic warrants a book-length treatment on its own, so I will not enter into discussions on geometry or other fields of mathematics, but at times I will point out connections to them. It should also be added that, when it comes to arithmetic, with the exception of Section 5.3, this book is focused on very basic arithmetic like addition and multiplication. There are two reasons for that choice. First, the scope of the book had to be limited somehow and, as interesting as pursuing the development of more advanced arithmetic would have been, it was not possible to tackle it here. Second, and more importantly, I am confident that the most important philosophical problems are present already when discussing basic arithmetic. When they are not, as in the topics of formalising arithmetic and infinity, I have expanded the scope.

I.1 A Priori Philosophy of Mathematics

I.1.1 Kant's Synthetic A Priori Knowledge

A good place to start unravelling the cluster of problems involved in explaining the character of mathematical knowledge is the traditional epistemological distinction between a priori and a posteriori. The distinction itself is older, but it is most commonly associated with the work of Immanuel Kant (1787) in *Critique of Pure Reason*.[1] The fundamental distinction is that a priori knowledge is acquired independently of experience while a posteriori knowledge is obtained through experience. With this distinction in place, further distinctions can be made. A priori knowledge, since it does not depend on our experiences, was standardly considered to be both *necessary* and *universal*, while a posteriori knowledge was considered to never be necessary or universal.[2] To complete this famous Kantian framework in epistemology, we need one last distinction, the one between *analytic* and *synthetic* propositions. According to this semantic distinction, analytic propositions are true entirely by virtue of their meaning, while the truth of synthetic propositions depends on the world and their relation to it.

Some examples will help to make these distinctions clearer. The proposition expressed by the sentence 'the mass of the fifth planet from the sun is greater than the mass of the sixth planet from the sun' is synthetic.[3] There is nothing in the meaning of the sentence that dictates whether it is true or false; that depends on how things stand in the world. As it happens, scientists have established that the proposition is true, based on observations of the solar system. These observations are complex in character. They require instruments like telescopes, as well as mathematical calculations. A theory of gravitation, at present either Newtonian mechanics or the general theory of relativity, needs to be applied in order to calculate the masses of the two planets. These physical theories, in turn, rely heavily on mathematical theories. It is only with the help of these physical and theoretical instruments that we are able to establish the truth of the proposition. Nevertheless, knowledge of that truth clearly depends on experience. It is only through applying our senses, in this case primarily vision, that we are able to ascertain that Jupiter, the fifth planet, indeed has a greater mass than Saturn, the sixth planet. Therefore, the proposition is

[1] I acknowledge that my account of Kant's philosophy here is quite cursory, meant to illuminate some key philosophical notions concerning mathematics rather than engage in Kant-scholarship.

[2] These connections were accepted for a long time, but some more recent accounts of a priori question this traditional view (see, e.g., Casullo, 2003). In addition, as famously argued by Kripke (1980), it can be possible in some cases (e.g., 'water is H_2O') that also a posteriori knowledge is necessary.

[3] Here 'proposition' refers to the content of the English-language sentence. Since the formulation 'proposition expressed by the sentence' is quite clumsy, from here on I will use the terms 'proposition' and 'sentence', as well as their instances, synonymously.

knowable a posteriori. Moreover, since the truth of the proposition clearly depends on the world, it is a case of *synthetic a posteriori* knowledge.

This example should make it clearer what the differences between a priori and a posteriori, on the one hand, and between analytic and synthetic, on the other hand, entail. Since the above proposition about Jupiter and Saturn is synthetic a posteriori knowledge, under the traditional view it is neither necessary nor universal. That Jupiter has a greater mass than Saturn is a contingent fact about the solar system; with a different course of actions after the Big Bang the matter could have been otherwise. In addition, in another solar system it could be the case that the sixth planet is indeed bigger than the fifth planet from the sun, making the truth a non-universal one. The truth may *appear* to be as certain as any truth, yet it is categorically distinct from analytic necessary and universal truths.

What kind of truths are analytic? In Kant's (1787) view, in an analytic truth the concepts in the proposition are in a particular type of containment relationship. The proposition 'all bachelors are unmarried' is a standard example of this. It is true entirely in virtue of its meaning, that is, the definition of the subject 'bachelor' includes the predicate 'is unmarried'. Unlike in the case of Jupiter and Saturn, we do not need to observe the world to ascertain its truth.[4]

With the distinctions between a priori and a posteriori, as well as analytic and synthetic, in place, Kant asked what kind of knowledge there is. We have seen that there is synthetic a posteriori and analytic a priori knowledge. For Kant, analytic a posteriori was an impossible combination but, importantly, he argued that there is synthetic a priori knowledge, that is, knowledge that is acquired independently of observations but which is nevertheless true not only in virtue of the meaning of the propositions. Metaphysics, the kind that Kant was involved in, was one case of synthetic a priori knowledge. But, for the present purposes, the more interesting case is *mathematical* knowledge, which Kant also saw as being synthetic a priori. Kant used examples from both arithmetic and geometry to make his point. In this book I will focus on the former idea of Kant, that is, that both syntheticity and apriority are characteristics of arithmetical knowledge. In the sentence '7 + 5 = 12', for example, the meaning of '12' is not included in the meaning of '7' or '5', or the combination of the two

[4] Although we clearly need to use our senses *somehow* to establish the truth of the proposition, simply to learn the meanings of the words. But after we have learned them, we need no further observations to determine that the sentence is true.

(Kant, 1787, B15–B16). But nevertheless, this knowledge is a priori in character. It is acquired by reason, and not through observation. Thus, Kant argued that arithmetic, like all pure mathematics, is synthetic a priori, while also being necessary and universal.[5]

I.1.2 Mathematical Objects and Platonism

In that last claim of Kant, however, resides a potential problem. If arithmetical knowledge, as part of pure mathematics, is synthetic, how can it be necessary and universal? If a proposition being synthetic means, as specified in Section I.1.1, that its truth depends on its relationship with the world, does this not imply that propositions of pure mathematics should be assessed in terms of their relationship with the world of mathematical objects? As it has turned out, answers to this question have dominated the philosophy of mathematics ever since Kant. In addition, pre-Kantian philosophical views concerning mathematics can also be framed in terms of this question.

Although it was present already in pre-Socratic philosophy, the most famous and influential pre-Kantian view concerning mathematics can be found in the work of Plato. He argued that, since sense perception can never give knowledge about the mathematical objects themselves, knowledge about mathematical objects must be gained through reason and recollection (Plato, 1992, 527a–b). Thus, the idea of mathematical knowledge being a priori was already present in antiquity. But equally crucial to Plato's philosophy was the idea that there is a world of mathematical objects and by reason we can gain knowledge of that world. In the philosophy of mathematics, the realist position according to which mathematical objects and structures exist in a mind-independent fashion is still called *platonism*.[6]

[5] Although for the present purposes this account is sufficient, it should be noted that the analyticity and syntheticity of mathematical knowledge has been a much-discussed topic in the literature (see, e.g., Boghossian, 1997).

[6] I follow here the custom that Platonism with a capital 'P' refers specifically to Plato's philosophy, while platonism with a lower case 'p' refers to a more general realist metaphysical position on mathematics (see Balaguer, 2016). Tait (2001) has suggested that instead of platonism, it would be clearer to talk about 'realism'. In the name of terminological congruity with the relevant literature, I adhere to the more common custom. Following another common custom in the literature (e.g., Dummett, 2006), I use the term 'mind-independent' to refer to something that is not dependent on human conventions, practices, languages and thoughts. Interestingly, recently a case has been made by Landry (2023) that Plato was in fact not a mathematical platonist, which would make the distinction between Platonism and platonism all the more crucial.

Platonism clearly provides an answer to the question about syntheticity. Propositions of pure arithmetic are synthetic because their truth depends on the state of things in the platonic world of mathematical ideas. Since truths about that world are necessary and universal, arithmetical knowledge can be at the same time synthetic a priori and both necessary and universal. The problem with this answer, however, is in explaining how our mental faculties can be used to get knowledge of mathematical objects in the platonic world. If mathematical objects such as numbers exist in the platonist sense, they must be *abstract*. But abstract objects, by definition, are non-spatial, non-temporal and causally inactive. This causes a problem in combining our cognitive faculties with the abstract world of mathematical objects. As famously asked by Benacerraf (1973), how can we as physical subjects get knowledge of abstract, non-physical objects? Some platonists, like Gödel (1983) and Penrose (1989), have proposed a special epistemic faculty for mathematics as an answer. However, in modern philosophy of mathematics, such explanations have found limited popularity. Instead, as we will see in Section I.1.3, modern platonism has tended to deal with this epistemological problem by distancing itself from Kant's idea that mathematical knowledge is synthetic a priori.

I.1.3 Analytic A Priori and Mathematical Objects

The importance of Benacerraf's (1973) question for modern philosophy of mathematics is indisputable. By pointing out the epistemological difficulty of combining an empiricist, observation-based, epistemology with a platonist ontology of mathematical objects, his question is problematic also for the Kantian notion of synthetic a priori mathematical knowledge. After all, if not the platonic world of mathematical ideas, then some other mind-independent aspects must determine mathematical truths – otherwise mathematical knowledge would be *analytic* a priori. But what could these aspects be and how could we have epistemological access to them?

In Section I.1.4, I will propose an answer to this question which I will defend throughout this book. But in the contemporary literature on philosophy of mathematics, especially in the tradition following Frege (1884), a more common solution to the problem is to deny that mathematical knowledge is synthetic a priori and contend that it is analytic. However, this rejection prompts a potential difficulty. If mathematical truths are analytic, how can we include mathematical *objects* in our mathematical theories? Truths like 'all bachelors are unmarried' are unproblematic analytic truths since they do not make the claim that there exist such things as bachelors. The only claim that is made is that *if* somebody is a bachelor, then he is unmarried. Certainly, many mathematical statements are like that, for example, the arithmetical statement $\forall n(n = n)$ ('all numbers are equal to themselves'). However, integral to mathematics are *existential* statements, for example, of the type $\exists n(3 < n < 5)$. For that statement to be true, there has to exist a number between 3 and 5. But how can we make an analytic existential statement like that? Does the truth of the statement not depend on whether numbers exist or not? And if numbers do not exist, as Hartry Field (1980), among others, has argued, such existential statements are, strictly speaking, false. Hence, a great part of mathematical statements, given a literal reading, would not be analytic truths but, more than that, they would not be truths *at all.*

A famous answer to this problem was provided by Frege (1884), who argued that also analytic a priori truths can make existential claims, through introducing new objects. In the following analytic truth, for example, we introduce new abstract objects:

> The direction of line A is the same as the direction of line B if and only if A and B are parallel.

Through the equivalence relation of lines being parallel, we are able to introduce new abstract objects, namely directions, by an analytic truth. Similarly, it is possible to introduce numbers as abstract objects. An analytic truth called *Hume's Principle* (Boolos, 1998, p. 181) states that the number of Fs is equal to the number of Gs if and only if the Fs and the Gs are equinumerous, that is, there is a one-to-one correspondence between the Fs and the Gs:

$$\#F = \#G \leftrightarrow F \approx G.$$

Here the operator '#' takes the concept of one-to-one correspondence and introduces new objects, that is, natural numbers.

These two analytic truths are cases of *abstraction principles* (see, e.g., Linnebo, 2018b). They 'carve up' previous propositional conceptual content (e.g., about equinumerosity) and give us a criterion of identity for new objects (e.g., natural numbers). Such abstraction principles have been used to refer to abstract mathematical objects by the neo-Fregeans Bob Hale and Crispin Wright (2001, 2009), as well as recently by both Agustín Rayo (2015) and Øystein Linnebo (2018b). Rayo calls his position *subtle platonism,* while Linnebo writes about 'thin objects'. The underlying idea in both accounts is that, by introducing mathematical objects via abstraction principles, we avoid Benacerraf's (1973) problem because the existence of abstract objects does not make any demands of the world. There is nothing problematic about epistemic access to abstract objects, since they are simply introduced by analytic truths, more specifically as singular referents in abstraction principles like Hume's Principle (for more, see Pantsar, 2021c). Thus, according to Linnebo, through such a process of 'reconceptualisation' we get ontological commitment to objects such as numbers (Linnebo, 2018b, p. 23). We can therefore make existential statements about mathematical objects without being inconsistent, all the while staying within the realm of analytic a priori.

I.1.4 Kant Revisited and Reinterpreted

We will return to the topic of the previous section in Chapter 9, but for now it is important to recognise that the neo-Fregeans, Rayo and Linnebo all appear to be committed to the view that mathematical knowledge is analytic a priori. Indeed, the Kantian notion of mathematical knowledge as synthetic a priori has become increasingly unpopular in modern philosophy of mathematics. However, I believe that this is potentially misguided. As I will show in this book, there is a lot to like about the idea that mathematical knowledge is synthetic a priori, as long as we can explain properly what we mean by 'synthetic' and 'a priori'. The way I will argue for this position is by including state-of-the-art data and theories from the empirical study of arithmetical and early quantitative cognition. Naturally, none of this data were available for Kant (or, for that matter, Frege). But I contend nevertheless that modern empirical research can be fruitfully understood in a Kantian context.

One of the most famous parts about Kant's *Critique of Pure Reason* is the way he compares his philosophy to the revolution in astronomy due to Copernicus:

> Hitherto it has been assumed that all our knowledge must conform to objects. But all attempts to extend our knowledge of objects by establishing something in regard to them *a priori*, by means of concepts, have, on this assumption, ended in failure. We must therefore make trial whether we may not have more success in the tasks of metaphysics, if we suppose that objects must conform to our knowledge. (Kant, 1787, Bxvi)

What I will be arguing for in this book is something similar: I propose that we can understand the nature of mathematical objects better if we focus primarily on the characteristics of mathematical knowledge. In this, I am moving along the same lines as Rayo (2015) and Linnebo (2018b) who also stress the primariness of knowledge (for them, concerning true abstraction principles) over objects in their argumentative methodology. However, there is an important difference in my approach. While I agree that, under a suitable (contextual) definition, mathematical knowledge is a priori, I do not believe that traditional a priori philosophical methodology suffices, or should even be prioritised, in studying the character of mathematical knowledge.

I.2 A Posteriori Approaches to Mathematical Knowledge

I.2.1 Empiricism

So far, I have focused on a priori approaches to mathematical knowledge, but there have also been well-known a posteriori approaches. These should be divided into two distinct positions. According to the first one, empirical research is integral to understanding what mathematical knowledge is like and how it is acquired. Let us call this the *methodological* a posteriori approach. According to the second position, arithmetical knowledge is fundamentally a posteriori. Let us call this the *epistemological* a posteriori position.

One can simultaneously subscribe to both the methodological and the epistemological a posteriori views on mathematical knowledge, that is, hold the position that mathematical knowledge is fundamentally based on observations and that it should be studied via empirical methodology. But it is also possible to subscribe to just one of the a posteriori views. Indeed, the account I argue for in this book takes empirical methodology to be crucial for understanding the development and character of arithmetical knowledge. Yet, I will argue that arithmetical knowledge is best understood as a priori, under a proper understanding of the term.

But this connection between methodological and epistemological a posteriori can also go the other way around. Historically, this other direction has in fact been more common. The most famous accounts that connect mathematical knowledge to the empirical, those provided by John Stuart Mill (1843) and Philip Kitcher (1983), both stick to essentially similar a priori philosophical methodology, as have platonist philosophers traditionally. Both accounts are empiricist in that they take mathematical knowledge to be fundamentally empirical in character, yet their methodology does not involve significantly engaging with empirical studies on mathematical cognition or mathematical practice. For Mill, the reason could be simple: there simply was no reliable and relevant empirical data available in the nineteenth century. And, even though Kitcher wrote his book almost a century and a half later, the situation had changed surprisingly little. As will be seen, there was already empirical research that pointed to future directions in the study of arithmetical cognition, but in the early 1980s the situation was radically different from the current state of the art. Therefore, the late emergence of methodological a posteriori approaches to mathematical knowledge is perhaps best explained by the simple fact that reliable and relevant empirical research has only started to accumulate during the last few decades.

But why was the epistemological a posteriori approach so unpopular among philosophers for so long – and indeed still is? Given that generally in epistemology there has been less and less room for rationalist approaches since the eighteenth century – although with notable exceptions like Hegelian philosophy – this was a somewhat peculiar state of affairs. As empirical data was given increasing importance in epistemology, mathematical knowledge remained largely a bastion of rationalism. The way mathematical knowledge was elucidated was through the traditional a priori methodology, with the – either explicit or implicit – understanding that mathematical knowledge is fundamentally different from most other types of knowledge.

The reason for the historical lack of popularity of epistemological a posteriori accounts can, I believe, be elucidated by an analysis of Mill's empiricist philosophy in his main work on the topic, the book *A System of Logic* (Mill, 1843). Previous empiricists like Hume had held mathematical truths to be analytic, but in Mill's philosophy mathematical statements are not given any special epistemological status. For him, mathematical knowledge – including that of arithmetic – is based on experiences of the physical world, just like all knowledge. The necessity of even the very basic laws, such as the law of identity $\forall x(x = x)$ is questioned by Mill: 'How can we know that a forty-horse power is always equal to itself, unless we assume that all horses are of equal strength?' (Mill, 1843, p. 170). Since all horses quite clearly are not of equal strength, Mill seems to argue that we cannot know that a forty-horsepower is always equal to itself. But this line of reasoning is quite confusing. If Mill means that we cannot equate the strength of an actual group of forty horses with that of another, he is of course correct. But it is hard to see the point of this observation, since nobody would claim the opposite. Similarly, if he means that we cannot be sure that the abstract unit of power 'forty-horsepower' does not match the power of all groups of forty horses, he is no doubt making a correct claim, but again nobody would claim the opposite. But if Mill's point really is to question that the abstract unit of 'forty-horsepower' is equal to itself, he must be taking mathematical objects to be tied to physical objects in a very concrete way (for more, see Pantsar, 2016a).

For Frege, Mill's philosophy meant throwing away everything that is precious about mathematical knowledge, and he ridiculed Mill's view as reducing mathematics to 'pebble arithmetic':

> Often it is only after immense intellectual effort, which may have continued over centuries, that humanity at last succeeds in achieving knowledge of a concept in its pure form, in stripping off the irrelevant accretions which veil it from the eyes of the mind. What, then, are we to say of those who, instead of advancing this work where it is not yet completed, despise it, and betake themselves to the nursery, or bury themselves in the remotest conceivable periods of human evolution, there to discover, like John Stuart Mill, some gingerbread or pebble arithmetic. . . . A procedure like this is surely the very reverse of rational, and as unmathematical, at any rate, as it could well be. (Frege, 1884, p. vii)

Frege was thus worried that Mill's effort to found arithmetic on empirical origins will take us away from the true essence of arithmetical truths and move the discourse to irrelevant details concerning the historical or personal discoveries of the truths. Given the direct connection Mill saw with physical objects (like horses) and abstractions (like horsepower), this criticism is easy to understand. Nevertheless, the origins that Frege saw as irrelevant are irrelevant to him because of his prior philosophical conviction that mathematical propositions have an essence that is independent of the cognitive origins of understanding them. Therefore, in the above passage Frege seems – to some degree – to conflate mathematics and the *philosophy* of mathematics. Clearly in mathematics, we do not want to bring in the empirical and psychological dimensions that may be involved. But, if we dismiss those dimensions as irrelevant for the epistemology of mathematics, we have already – without argument – rejected the view that mathematical knowledge could be somehow based on an empirical foundation.[7]

From a modern perspective, it is not as easy to accept Frege's dismissal of the possibility of mathematical knowledge having an empirical foundation. Mill's empirical account may have been crude in its understanding of abstraction, but a more sophisticated empiricist approach to mathematics was presented by Kitcher (1983). In Kitcher's view, mathematical

[7] For more general considerations on the topic of bringing psychological factors into epistemology, see Kitcher (1992).

knowledge concerns generalisations of operations that we do in our environment. In our environment, we may play around with collections of pebbles and learn that the operations obey certain rules. We can then generalise on these rules and end up with rules of arithmetic applicable not only to small collections of pebbles, but also to larger quantities of any kind of objects, even up to infinity. Frege might continue to dismiss this as 'pebble arithmetic', but there is an important difference to Mill's empiricist philosophy. The pebbles (or other physical objects) play an integral psychological role in the process, but the arithmetical rules we learn in this way are *operational generalisations*, that is, abstractions that are no longer tied to particular sets of objects.

This topic will be given a lot of attention throughout this book, but initially it seems plausible that Kitcher is fundamentally correct about the way we first learn about arithmetical rules. At least for most people, empirical aspects play an important role in the early learning of arithmetic. As we will see, it is also plausible that the development of arithmetic as an intellectual discipline was initially tightly connected to empirical aspects. However, we can accept these origins without making the empiricist assumption that arithmetical knowledge is a posteriori. There are two reasons for this. First, after being aided by physical objects in the learning process, the empirical methods are *not indispensable*, which makes arithmetical knowledge different in character from empirical knowledge. Children may use their fingers to count and add, for example, but, while they may be integral for the learning process, such empirical methods can be abandoned later. Second, even if arithmetical statements had their origins in empirical procedures, they are not empirically confirmable or falsifiable under any relevant reading of 'empirical'. For one thing, arithmetical calculations are neither corroborated nor refuted by direct experiment. If some arithmetical theorem is proved, an empirical corroboration of the proof has no value. The sum $4,289,031 + 9,802,472 = 14,091,503$ is true because it follows the axioms of arithmetic. Conducting a physical experiment with, say, pebbles would not make this any more certain.

It is thus crucial to note that mathematical proof has its own special character and an epistemological theory must be sensitive to that. Hence, empiricist theories of mathematical knowledge seem to be fundamentally flawed, even though I – unlike Frege – support the idea that the empirical processes that are used in learning arithmetic are relevant for the epistemology of arithmetic. What I agree with Frege on is that these processes should not be confused with the character of arithmetical *truths*. In modern literature, this distinction is best known as the difference between the context of *discovery* and the context of *justification* (Frege, 1879). What Frege is writing about is the context of justification, how arithmetical truths can be *established*. He deems the context of discovery, how arithmetical truths are *learned*, to be irrelevant for the context of justification. This was the foundation of Frege's criticism toward psychologism of the late nineteenth century, in particular that of Schröder (1873).

What Frege thus wanted to do was to find a logical basis for arithmetic that is independent of the unpredictability of the psychological processes involved. Frege attributes to Schröder the view that we arrive at the concept of number by abstracting away every other property, such as colour and shape, from our observations. When we see, for example, a bag of oranges, we can abstract away everything that distinguishes the objects as oranges and arrive at the number of oranges. Frege (1884) pointed out serious difficulties with this approach. For instance, there are often various ways of extracting a number out of objects. If we have, say, a plateful of grapes, do we mean the number of individual grapes, the number of bunches of grapes, or perhaps something else – that there is one plate, for example? For Frege, a number could not be a property of the objects because our choice of extracting number is arbitrary in a way that the colour of the grapes, for example, clearly is not.

In general, I believe Frege's criticism of psychologism mostly hits its target (see Pantsar, 2016a for more). Like Frege, I find abstraction of the Schröder-type to run into the danger of arbitrariness: only certain types of abstraction let us extract the desired number, but in Schröder's account there does not seem to be any reason why one would prefer one extracted number over another. However, a lot has happened in the century and a half after Schröder presented his theories. As I will argue in this book, currently we have good reasons to think that a certain type of abstraction to the numerosity of macro-level objects *does* have a privileged position over other types. But even so, I will show, it is problematic to conceive of number as a property of objects. Instead, what I will do in this book is to follow Kant's Copernican revolution. We should not focus primarily on what the objects that numbers are associated with are like. Instead, we should focus on what *knowledge* of numbers is like and how it is made possible by cognitive processes.

I.2.2 Empirical Study of Mathematical Cognition

While I have criticised above the epistemological a posteriori approaches to mathematical knowledge, in this book I will make substantial use of the *methodological* a posteriori approach. Indeed, one of the main tenets of this book is that the philosophical study of mathematical knowledge should be informed by the empirical study of mathematical cognition and its development. Mathematical cognition is a relatively new field for empirical study and solid results have only properly emerged during the past few decades. This is not to say that the empirical study of mathematical cognition does not have a longer history. From learning mathematics (e.g., Skemp, 1987) to mathematical invention (e.g., Hadamard, 1954), mathematical cognition was an important topic in psychology in the latter part of the twentieth century. While much of that work also carries philosophical relevance, for the epistemology of arithmetic the important change has taken place mostly starting from the 1990s. Since then, a widespread view has started to emerge that arithmetical ability and arithmetical knowledge are not solely the domain of sufficiently mature, educated human individuals. Data emerged on infant ability with numerosities, most famously in the experiment reported by Karen Wynn (1992), according to which infants can carry out rudimentary addition and subtraction operations. Similar arithmetical abilities were reported in many animals, including 'numerical and arithmetic abilities in non-primate species' (Agrillo, 2015) and 'arithmetic in newborn chicks' (Rugani et al., 2009).

Stanislas Dehaene, one of the most influential researchers in the field, argued in his widely read book *The Number Sense* (Dehaene, 2011) that these abilities are due to our brains possessing an innate number sense that we share with many non-human animals:

> How can a 5-month-old baby know that 1 plus 1 equals 2? How is it possible for animals without language, such as chimpanzees, rats, and pigeons, to have some knowledge of elementary arithmetic? My hypothesis is that the answers to all these questions must be sought at a single source: the structure of our brain. (Dehaene, 2011, p. xvii)

For the epistemology of arithmetic, the verification of Dehaene's hypothesis would be monumental. Instead of looking for explanations of arithmetical knowledge in mathematical objects, we should focus on the structure of the brain. If arithmetical knowledge, as Dehaene claims, is present in babies and non-human animals, then clearly abilities exclusive to sufficiently mature humans, such as language, are not necessary for the development of arithmetic. This would suggest a Kantian turn with a modern empirical twist: if we can explain how arithmetical knowledge emerges in non-human animals and human infants, we are likely to be dealing with the very foundation of arithmetical knowledge. Even modern formal axiomatic systems would essentially be expansions and specifications of an innate arithmetical cognitive system that is entirely the product of biological evolution.

Susan Carey, another researcher whose work will be extensively discussed in this book, has called these types of evolutionarily developed cognitive systems *core cognition* (Carey, 2009).[8] Core cognitive systems are thought to differ 'systematically from both sensory/perceptual representational systems and theoretical conceptual knowledge' (Carey, 2009, p. 10) and core cognition is thought to be 'the developmental foundation of human conceptual understanding.' (p. 11). In Chapter 1, we will see what kind of systems have been suggested as the relevant core cognitive systems for the development of arithmetical knowledge. But, for now, it is important to emphasise the difference of the core cognitive framework to the kind of empirical work that was done on mathematical cognition for most of the twentieth century. The psychology of learning mathematics was an important topic and Piaget's (1970) theory of stages of cognitive development, for example, had a great influence on mathematics education (see, e.g., Ojose, 2008). But, contrary to Piaget, who believed that arithmetical knowledge is only possible once the child has gone through several stages of cognitive development (namely, the stages of *concrete operations* and *formal operations*, after going through the earlier *sensorimotor* and *preoperational* stages (Piaget, 1970)), Dehaene is arguing that already infants only a few months of age have arithmetical knowledge. This is a paradigmatic difference when it comes to applying empirical data to the study of arithmetical knowledge.

I.3 A Synthesis of Two Approaches

I.3.1 Incongruent and Misleading Terminology

In this book, I will argue that arithmetical knowledge is best studied in an interdisciplinary manner, using methodology and results from both the empirical study of numerical cognition and the philosophical study of arithmetical knowledge. To be more precise, rather than explicitly arguing for this position, I will show that applying this interdisciplinary approach

[8] This is a modification of Elizabeth Spelke's term 'core knowledge' (e.g., Spelke & Kinzler, 2007). As Carey points out, representations in core cognition need not constitute knowledge, so 'core cognition' is definitely a more fitting term. Indeed, as we will see, many confusions could have been avoided with stricter use of the term 'knowledge'.

gives us the best understanding of arithmetical knowledge and how it is developed and acquired. One important problem in establishing a platform for fruitful interdisciplinary research, however, is to agree upon congruent terminology. This is a particularly pressing challenge in the empirical research on numerical cognition since the literature in different disciplines includes radically different application of the central terms. In the noted quotation of Dehaene, for example, he asks 'How can a 5-month-old baby know that 1 plus 1 equals 2?' But, from a philosophical perspective, this question makes little sense. Whatever the ability of the baby is, few philosophers would be ready to ascribe *knowledge* to it.

Yet it would be unreasonable to dismiss Dehaene's question just because his language does not conform to the standard jargon of epistemologists. Whatever he means by knowledge, we can be confident that Dehaene does not think that infants have arithmetical knowledge in the same sense as mature humans have knowledge of the statement $1 + 1 = 2$. In this way, I believe philosophers should generally give empirical literature a charitable reading and not get stuck with terminological issues, just as empirical scientists should give philosophers the benefit of doubt when they show lack of fluency with the empirical literature.

However, this kind of application of the principle of charity does not render the question of what exactly Dehaene *did* mean any less important. It is my belief that not only interdisciplinary research with philosophers but also empirical research itself would benefit from a clear and unambiguous use of terminology. Let us take a look at a couple of examples to see why. As mentioned earlier, one of the most influential experiments in the field of numerical cognition was conducted by Karen Wynn (1992). In the experiment, 5-month-old infants were shown dolls in settings that were designed to test whether the infants had numerical abilities. The infants were shown one doll and another doll put behind a screen. In some trials, one of the dolls was removed clandestinely before the screen was lifted. In others, both dolls remained. In another variation, the infants were initially shown two dolls and then after one was visibly removed, either one or two dolls were revealed (Figure I.1).

What Wynn's experiment (and its many replications) showed was that the infants reacted with surprise (measured as longer looking times) to the situation in which only one doll remained. This is what Dehaene referred to when he asked how 5-month-old infants can know that one plus one equals two. In Wynn's formulation: 'Here I show that 5-month-old infants can calculate the results of simple arithmetical operations on small numbers of items. This indicates that infants possess true numerical

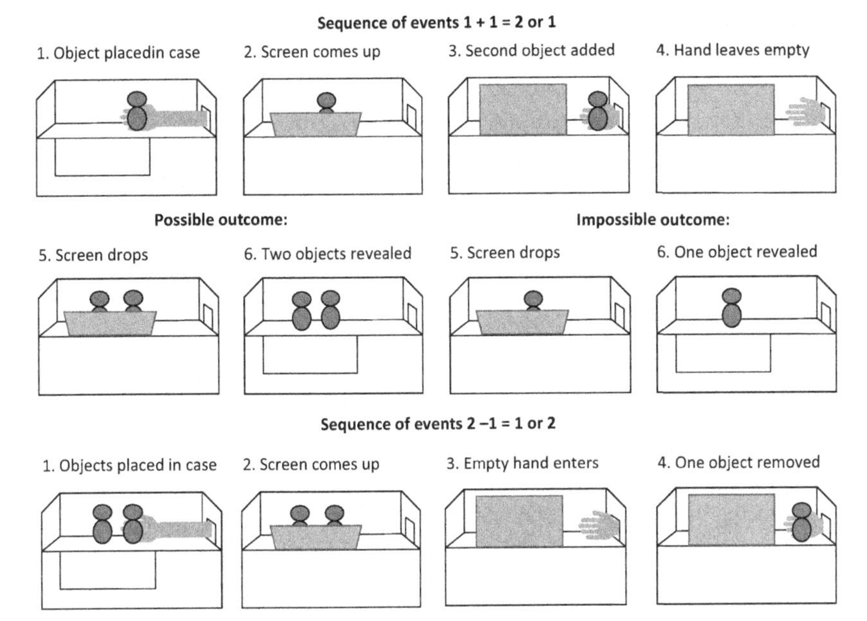

Figure I.1 The setting on the experiment reported in Wynn (1992).*
* She tested 5-month-old infants in trials that alternated between one-item and two-item results. The experimenter removed one doll through a hidden trap door in half the trials. The possible and impossible outcomes are reversed for the lower sequence of events.
After Bremner et al. (2017).

concepts, and suggests that humans are innately endowed with arithmetical abilities' (Wynn 1992, Abstract). Several claims are made in this short passage. Infants can *calculate* results of *arithmetical operations*. They possess *numerical concepts* and have *innate arithmetical abilities*. In this book, I show that all these statements are fundamentally wrong. Nevertheless, I believe that Wynn's research is highly relevant and the results applicable in explaining the development of arithmetical knowledge.[9]

The seemingly contradictory position in which I can take claims by empirical researchers to be wrong but still consider them important for explaining arithmetical knowledge becomes unproblematic when we understand the root of the problem. As will become clear in this book,

[9] It is important to note that there are several ways in which 'innate' can be understood, which has led Griffiths (2001), for example, to conclude that the entire notion should be replaced by more specific terms. In the present context, 'innate' refers generally to genetically determined, evolutionarily ancient capacities which may or may not be present at birth.

I do not believe that infants or non-human animals calculate results of arithmetical operations, nor do I believe that they possess innate numerical concepts or arithmetical abilities. What they do is show behaviour that can be *described* in terms of arithmetical operations. The infant does not know that one plus one equals two, but if it *did* know it, the behaviour would likely be quite similar. Thus, the problem is not in the experimental data itself, it is with the cognitive ascriptions made in order to explain the data.

One important consequence of this type of use of incongruent terminology, as pointed out by Rafael Núñez (2009), is the potential confusion of the *explanans* with the *explanandum*:

> A major problem in most accounts of the concept of number is that scholars often introduce crucial elements of the explanans in the very explanandum. That is, they take number systems as pre-given and introduce them as a part of the explanatory proposal itself ... Gallistel et al. (2006, 247), for instance, speak of 'mental magnitudes' referring to a 'real number system in the brain', where the very real numbers are taken for granted, and put them 'in the brain'. (Núñez, 2009, p. 72)

The task at hand for Gallistel et al. (2006) was to explain the development of real number systems based on an evolutionarily developed mental magnitude system. In Section 1.2, we will see how such explanations are developed in detail, but it is easy to see that something strange is going on if the mental magnitudes are equated with real numbers. The former, according to the hypothesis of Gallistel and colleagues, are innate and the product of biological evolution. The latter, on the other hand, are the product of a long line of development in mathematics. Perhaps there could be a straightforward explanation of how real number concepts are developed and acquired based on innate mental magnitudes – although I will argue that this is not the case – but we must not confuse the two notions (real number concepts and innate mental magnitudes) when presenting the phenomenon that needs to be explained.

These types of problems concern many accounts of the cognitive basis of numbers. When philosophers discuss numbers, they are concerned with abstract objects. There are many different ways of understanding what this abstractness entails (see, e.g., Rosen, 2020 for details), but the general understanding is that numbers as abstract objects should be distinguished from mental or physical *representations* of numbers. Whatever numbers are, they are something different from my utterance of the word 'three' or writing the symbol '3' on a piece of paper. Importantly, the abstractness of numbers also makes it necessary to distinguish between numbers and mental number *concepts*. At some point in my development, I acquired the number concept THREE.[10] However, this concept is not the *number* three. This is a crucial point to make, and it goes against the terminology used by many empirical researchers. Gallistel, for example, writes that 'Behavioural and electrophysiological data now convincingly establish the existence of numbers in the brain – in animals from insects to humans' (Gallistel, 2018, p. 1). Ironically, Gallistel had earlier, together with Gelman, proposed the term 'numeron' to refer to the representation of 'number in the brain' (Gelman & Gallistel, 1978). But now that 'numbers' and 'numerons' are confused, we get sentences like the following: 'Much of the work on numbers in the brain focuses only on the referential role of numerons – more particularly, on their reference to numerosities' (Gallistel, 2018, p. 1). I agree with Gallistel about the content of much of the work that he is referring to. The reference of numerons to numerosities is indeed a central topic of research for the topic at hand. However, that work is *not* about numbers in the brain. Granted, it is possible to entertain the view that numbers exist in the brain, and only in the brain. But that is a specific – and quite radical – claim that cannot simply be assumed to hold.

I.3.2 Key Terminology, Conceptual Analysis and Empirical Data

To avoid the kind of confusions discussed in Section I.3.1, it is clear that we must get rid of the incongruent terminology. To begin with, I want to establish a few basic terminological distinctions. This is not a new concern in the history of the study of numerical cognition. As detailed by Dos Santos (2022), already Stevens (1939) pointed out that specific terminology should be introduced for number-related concepts in the psychophysicist context. Davis and Pérusse (1988) asked for similar conceptual clarity in describing numerical competence in animals. Recently, both cognitive scientists (Núñez, 2017) and philosophers (De Cruz et al., 2010; Dos Santos, 2022; Pantsar, 2014, 2021a) have emphasised the need for distinct terminology for numbers as used in mathematics and quantity notions in the context of non-mathematical capacities.

In this book, I use a particular conceptual distinction aimed at avoiding the problems just referred to. Here *numerosity*, as used by Gallistel, is a general term for quantities. Infants, non-human animals and adult humans all observe their environments in terms of numerosities. This is what the behavioural and electrophysiological data referred to by Gallistel (2018)

[10] I follow here the Fodorian custom of referring to concepts in capital letters.

strongly establishes. *Numbers* are a specific type of numerosity. They form the abstract subject matter for arithmetic (in case of natural numbers), as well as other fields of mathematics (in case of calculus, complex analysis, etc.). They can be referred to by *numeral symbols* ($1, 2, 3, IV, V, 1/2, \pi, \sqrt{2}$, etc.) or by *numeral words* (one, two, drei, vier, cinque, etc.). In the brain, they are represented by *number concepts* (ONE, TWO, ONE THIRD, SQUARE ROOT OF TWO, etc.).

In this book, the subject matter is the arithmetic of natural numbers $0, 1, 2, 3, 4, \ldots, n, \ldots$ In mathematics, sometimes the number 0 is included in the natural numbers but sometimes it is not. When this distinction is important, it will be noted, but mostly we will be concerned with the set of *positive integers*, thus not including zero. The infinite set of natural numbers in mathematics is referred to by the symbol N. Natural numbers can be used both for counting (one, two, three, ...) and ordering (first, second, third, ...). In the first case, we speak of *cardinal* numbers. In the second case, we speak of *ordinal* numbers. Cardinal and ordinal numbers have very different properties when considering infinite sets, but for finite sets we can switch from one to another: the cardinal number four refers to the fourth ordinal number on the counting list and so on (this in case that natural numbers are considered to begin from 1). Therefore, each non-zero natural number is connected in most languages to two numeral words. As for numeral symbols, practices vary across languages. In English, it is common in some contexts to use Roman numeral symbols to refer to ordinal numbers (e.g., Elizabeth II, Super Bowl LVI, *Godfather III*). Indo-Arabic numeral symbols are used both for cardinal numbers and ordinal numbers, with ordinality expressed by either a dot or another suffix (e.g., 9th) after the numeral symbol.[11]

It is important to make the distinction between numbers, number concepts, numeral words and numeral symbols, as well as that between the cardinal and ordinal interpretation of natural numbers. While these distinctions are generally accepted, I believe that it is equally important to make similar distinctions when discussing the processing and manipulation of numbers and numerosities. With such distinctions in place, many confusions can be resolved. What goes wrong, for example, in Wynn's ascription of cognitive abilities to explain infant behaviour? As I understand it, the problem is at least partly rooted in the terminology and the conceptual distinctions that it reflects. The writings about 'infant arithmetic' and 'animal arithmetic' create an accepted use of language in

[11] The Indo-Arabic numeral system is often also called *Hindu-Arabic*.

which any treatment of quantitative information is called arithmetic. But this kind of usage is almost certain to run into ambiguity and downright confusion. Certainly, there is no reason to think that researchers like Wynn and Dehaene want to equate the infant or animal abilities with the arithmetical ability of mature humans. But, nevertheless, they choose to write about both abilities as 'arithmetic'.

Here I want to be emphatic about taking a different approach. *Arithmetic*, according to the best current knowledge, is exclusively the domain of sufficiently mature human subjects.[12] It involves an extensive grasp of exact number concepts and well-defined operations on them. Thus, arithmetic – to the best of our current knowledge – requires linguistic ability and is made possible in the individual development through processes of enculturation (Fabry, 2020; Fabry & Pantsar, 2021; Menary, 2015; Pantsar, 2014, 2018a, 2019b, 2020). Once we understand arithmetic in this sense, it is much easier to take stock of the different types of quantitative abilities. *Formal arithmetic* is what advanced and expert mathematicians are able to do. That includes proving arithmetical theorems and other knowledge and skills not shared generally by humans. But almost all sufficiently mature humans growing up in arithmetical cultures have some arithmetical knowledge.[13] They can count items and conduct basic arithmetical operations like addition, subtraction, multiplication and division. They can do this with the help of cognitive tools like pen and paper and the abacus, but most of them can also do *mental arithmetic*, that is, carry out arithmetical operations for small numbers without using tools.

All those kinds of knowledge and skills are arithmetical, and none of them are possessed by non-human animals or infants. But there are also human cultures, such as the Amazonian peoples of Pirahã and Munduruku, in which none of those skills are generally possessed by their members – child or adult (Frank et al., 2008; Gordon, 2004; Pica et al., 2004). Arithmetical abilities are thus present only in arithmetical cultures, and all other types of abilities with numerosities should be distinguished from them. I have suggested previously that the quantitative skills shown by anumeric cultures, non-human animals and human children before a certain developmental stage should be called *proto-arithmetical*

[12] If, say, non-human primates would show capacity for proper arithmetical calculations, this claim would need to be adjusted. But there is nothing to suggest that in the extant data. The one currently feasible exception is artificial intelligence, which can acquire mathematical capacities through different means, for example, pattern recognition (see Pantsar, 2023a for more).

[13] One notable exception is formed by people with *acquired dyscalculia* who have lost all or some of their arithmetical ability through, for example, brain injury.

(Pantsar, 2014, 2019b). Rafael Núñez has proposed the word 'quantical' to refer to roughly the same abilities (Núñez, 2017). I prefer the word 'proto-arithmetical', because it already points to an important claim that I defend throughout this book: that the evolutionarily developed quantitative abilities are the basis on which arithmetical ability develops.

With the distinction between proto-arithmetical and arithmetical in place, we can specify the important conceptual clarification between *numbers* and *numerosities*. Let us speak of 'numbers' only when discussing the referents of *exact number concepts* in the context of arithmetic or other mathematical areas, while using the word 'numerosities' in connection to proto-arithmetical abilities (De Cruz et al., 2010; Pantsar, 2018a). Finally, we need a term to distinguish numbers and numerosities, now connected to cognitive abilities, from the quantities that they represent or enable processing. Let us call these simply *cardinalities*. Importantly, cardinalities should be distinguished from *cardinal numbers*. One can get information about cardinality of a collection without possessing cardinal number concepts. For example, by matching each fork with a knife while setting a table, one can establish that the cardinality of the forks is the same as the cardinality of the knives. No knowledge of numbers is required. However, to establish that the cardinality of a collection is some natural number, one needs to possess (cardinal) number concepts.[14]

At this point, we should also discuss the controversial issue of representations. As framed above, I take number concepts to be mental representations of numbers, which conforms to the standard use in the literature (see, e.g., Carey, 2009). However, I will speak of representations also when discussing cognitive processes involving proto-arithmetical abilities. This has become controversial in the philosophy of cognition and mind in recent years. Proponents of (radical) *enactivist* cognition argue that representations are only present in the mind once there are linguistic truth-telling practices in place (Gallagher, 2017; Hutto & Myin, 2013, 2017; Zahidi, 2021; Zahidi & Myin, 2016). Since infants and non-human animals clearly do not have such practices, the radical enactivist account would denounce talk of numerosity representations in infant and non-human animal minds. Here I do not want to commit to any particular view about representations. I believe that the account I develop can be understood both in terms of radical enactivism and accounts that allow for non-linguistic representation (see Pantsar, 2023c for a detailed treatment

[14] Are our first number concepts cardinal or ordinal? This is an important question in the literature on numerical cognition. I will return to it in Section 9.4.

of this topic). Since a choice of terminology must be made, I have decided to talk about representations in a more general sense that could include, for example, visual representations or some kind of coding schemes. In any case, this should not be problematic since I have made the clear conceptual distinction between number concepts (which the radical enactivists would also accept as involving representations) and proto-arithmetical numerosity representations (which they would not).

Now that the key terminology is defined and the necessary conceptual analysis conducted, I believe that we are in a much better place to understand the kind of empirical data discussed by Wynn and Dehaene. The five-month-old babies do not have arithmetical knowledge. They do not possess number concepts, nor do they conduct arithmetical operations. But they *do* possess proto-arithmetical abilities with numerosities which enable them to get information about the cardinalities of collections of items. Just what are these abilities and how have they developed? How are they used in acquiring properly arithmetical abilities? These will be the topics of Chapter 1. But already we can see a way forward. Nothing about the experimental data of Wynn, for example, needs to be dismissed on the grounds of introducing new, coherent terminology. With the new conceptual framework in place, we only need to reinterpret the data. Perhaps some of the data needs to be ultimately rejected because it is revealed as problematic in the face of new data. But, as I will show, by combining philosophical conceptual analysis with data from empirical research, in most cases we get a better understanding of the cognitive phenomena involved.

I.3.3 A Project of Interdisciplinary Synthesis

In the manner described above, this book is a project of interdisciplinary synthesis in the research relevant to the development of arithmetical knowledge. While I also use a priori philosophical methodology, I am not committed to the position that arithmetical knowledge is a priori in character, let alone that arithmetical knowledge can be elucidated only through a priori methodology. In modern parlance, I am not a classic 'armchair philosopher'. However, the kind of conceptual analysis I want to conduct here is not *descriptive*, in the sense that it 'produce[s] a systematic account of the general conceptual structure of which our daily practice shows us to have a tacit and unconscious mastery' (Strawson, 1992, p. 7). In referring to Strawson's view, Somogy Varga writes that:

> In the contemporary philosophical landscape, such a view of the tasks and methods of philosophical inquiry is becoming much less common, and major scientific fields of inquiry are now complemented by subdivisions of philosophy that specialize in investigating a range of questions pertinent to the subject matter. The success of cognitive science has surely been a motivating factor for philosophers to account for new findings and to adjust their theories, topics, and approaches. Philosophers investigating the mind now often draw on findings in the sciences of the mind, reaching conclusions based on empirically informed reflection instead of a priori methods. (Varga, 2019, pp. 1–2)

While Varga writes in this passage about the philosophy of mind, I can wholeheartedly agree with his views when it comes to the epistemology of arithmetic. That is why I refer in this book to a vast array of studies in the cognitive sciences, from animal behaviour to cognitive neuroscience. This is crucial for achieving true progress in determining how arithmetical knowledge has developed. Yet this focus on empirical studies of cognition does not mean that more 'traditional' philosophical research does not play an important role in my methodology. Conceptual analysis and other philosophical methods are crucial for developing an understanding of how empirical data should be interpreted and understood. One reason for this is that empirical data, while increasing both in quantity and quality, still tells us relatively little about the development of arithmetical knowledge. There are still large parts that must be filled in with conjectures and hypotheses, to be tested by further studies.

Another reason is that conceptual analysis can help us see logical or epistemological connections between concepts that may well turn out to indicate connections also in cognitive development. After all, as I will argue in detail in Part II of the book, one of my main tenets in this book is that arithmetic is a culturally developed phenomenon. The development of arithmetic is thus tied to the general cultural development of knowledge, in which arithmetic as a discipline has been connected to many other areas, including philosophy. The history of the development of arithmetic should be studied in tandem with the study of the cultures in which arithmetic has developed. What was the status of arithmetic in that particular culture? Who were taught arithmetic, and with what aims? What applications did arithmetic have? These, and many other questions, are important in explaining how different cultures have developed arithmetical knowledge. Thus, the interdisciplinary approach, in addition to philosophy and the cognitive sciences, should also include the history of mathematics, cultural anthropology, archaeology, as well as mathematics itself.

Studies from all these disciplines will play an important role in this book.[15]

But most of all, I believe that an interdisciplinary synthesis between philosophy and other disciplines is needed because it *works*. It is the way we can get the best possible understanding of the development of arithmetical knowledge. The task at hand, after all, is huge. If the state of the art is correct, the extent of our innate proto-arithmetical abilities is not significantly different from that possessed by a wide range of non-human animal species. But somehow from these origins we have managed to create incredibly complex arithmetical theories whose applications have transformed the world in which we live in fundamental ways. In order to explain this kind of development, all relevant and reliable data are welcome, regardless of the discipline. That, however, makes it all the more important that data from different fields are analysed and interpreted in a unified, coherent theoretical framework. I hope that this book will make a positive contribution to this development.

I.4 Structure of the Book

As already mentioned, this book is divided into three parts. Part I focuses on the ontogeny of arithmetical knowledge. In Chapter 1, I explain the evolutionarily developed proto-arithmetical abilities for processing numerosities that humans already possess as infants and share with many non-human animals. These abilities (subitising, estimating) are then contrasted with proper arithmetical abilities. In addition to presenting the empirical data that establish the existence of proto-arithmetical abilities, the purpose of the chapter is to make clear that there is an important distinction between these evolved abilities and proper arithmetic of natural numbers.

Chapter 2 is about the way number concepts are acquired on the basis of proto-arithmetical abilities. First, a distinction is made between different types of abilities (conceptual, pre-conceptual) with numerosities. Then

[15] Here it should be clarified that while I believe in the general notion of culture as important for the development of human practices and cognitive abilities, I do not promote the kind of culturalism that takes cultures to form closed entities. The question of what should count as a culture will ultimately remain unanswered in this book. Societies share practices and norms up to different degrees, as well as being heterogeneous in other ways. Moreover, individuals may identify as a member of several cultures, or perhaps none at all. All these questions are important, and not without importance for the cultural development of arithmetic. However, they go beyond the scope of this book. Here I simply presuppose that discourse on cultures can be advantageous over other forms of discourse on collectives (e.g., nations, families, etc.). Thus, culture is used in this book primarily as a unit of discourse. I hope that the reasons behind this choice will become clear in Part II of the book.

three prominent accounts in the empirical literature are discussed in detail: nativism, Dehaene's (2011) 'number sense' and Carey's (2009) bootstrapping account. The bootstrapping account is deemed as the best way forward and it is then analysed in detail in light of critical views presented in the literature.

In Chapters 1 and 2, it is established that proto-arithmetical abilities should not be confused with acquiring number concepts and arithmetical abilities. The latter, rather than being of evolutionary origin, demand cultural learning. In Chapter 3, the framework of enculturation as proposed by Menary (2015) and others is presented in order to formulate a theory of number concept acquisition that is sensitive to both biological and cultural factors. This theory is then expanded to arithmetical skills and knowledge. The details of the enculturation framework are analysed carefully, including its neuronal realisation (i.e., neuronal recycling vs. neural reuse). Finally, empirical predictions of the enculturation account are formulated and discussed.

In Part II, the focus switches to phylogeny and cultural history; the way human abilities have evolved biologically and developed culturally. In Chapter 4, I move the analysis from the individual level to the phylogeny and cultural history of numbers and arithmetic on a cultural level. How do number concepts develop within cultures? I start from the following challenge posed by, among others, Pelland (2018): how could numerals emerge without there first being number concepts, as suggested by the enculturation account? I will explore the framework of material engagement of Malafouris (2013) as a possible answer to this, based on recent anthropological data. Then I discuss the best current knowledge on the cultural development of numeral systems, from the point of view of both historical and ethnographical studies. Finally, I will detail a theory, influenced by the ones presented by Wiese (2007) and dos Santos (2021), that numerals and number concepts have co-evolved through material engagement. I argue that this theory is consistent with the enculturation framework as it applies to ontogeny.

While Chapter 4 focuses on number concepts, in Chapter 5 the approach is extended to the cultural history of arithmetical skills. These range from basic arithmetic (addition, multiplication) to formal systems of arithmetic and infinity. The role of cognitive tools (writing systems, abacus, etc.), cognitive practices (e.g., enumeration, finger counting) and applications (bookkeeping, trade, agriculture, etc.) are discussed. It is shown that all stages of the development of arithmetic can be explained in the present approach.

In Chapter 6, a theoretical framework for this approach is detailed, based on accounts of *cumulative cultural evolution* as proposed by Henrich (2015), Heyes (2018), Boyd and Richerson (1985, 2005), and others. In incremental steps across generations, I will show, number concepts and arithmetical operations came to be through cultural learning. In virtue of the creativity and innovation of individuals and groups, as well as communication with other cultures, these developments may take new paths. Results from anthropology, ethnography and the history of mathematics are presented to support this account. Finally, the relation between enculturation (in ontogeny) and cumulative cultural evolution (in phylogeny and cultural history) is discussed, showing that these two accounts can be combined into one coherent theoretical framework.

Finally, in Part III, the considerations on arithmetical knowledge developed in Parts I and II are discussed in connection to the literature in the philosophy of mathematics. In Chapter 7, the topic is the epistemological importance of the first two parts of the book. First, the potential threat of conventionalism is discussed. I defend the view that the proto-arithmetical origins determine the trajectory of the development of arithmetical knowledge to a sufficiently strong degree to contradict strict forms of conventionalism. Nevertheless, conventions (e.g., symbols, words, practices) are important for both the ontogenetic and historical development of arithmetic. This is discussed also in terms of Wittgenstein's philosophy of mathematics, which has received different interpretations in the literature. Finally, I will argue that arithmetical knowledge, as described in Chapters 1–7 of this book, is best described as being *maximally inter-subjective*, that is, basic arithmetical knowledge is shared in a highly similar manner by most phenotypical members of all arithmetical cultures.

In Chapter 8, the account of arithmetical knowledge I have developed is analysed in terms of the classical characteristics of mathematical knowledge as described in traditional philosophy of mathematics: apriority, objectivity, necessity and universality. I show that each of these can be saved in my account in a sufficiently strong sense. Apriority should be interpreted as contextual apriority; the context being set by our proto-arithmetical abilities and their enculturated development. Objectivity should be understood as maximal inter-subjectivity, as detailed in Chapter 7. The necessity of arithmetic needs to be seen as the impossibility of strongly deviant arithmetical knowledge in arithmetical cultures. And finally, the universality of arithmetical knowledge should be understood as universality of arithmetic for phenotypical members of arithmetical cultures. For each notion, I will argue in detail how the adjusted interpretation is strong enough to save what is generally seen as being essential to arithmetical knowledge.

In the final chapter of Part III, Chapter 9, the epistemological considerations of Chapters 7 and 8 are analysed in terms of their ontological commitments. First, while the present account is compatible with traditional platonist views, according to which arithmetical objects have a mind-independent existence, no such ontologically heavy assumptions need to be made. Then I will discuss the accounts of thin objects (Linnebo) and subtle platonism (Rayo), which aim to provide ontologically 'light' versions of platonism. I will show that my account is compatible with both Linnebo and Rayo, but without the need to make any platonist assumptions. Finally, I will discuss the ontology of constructivism, according to which numbers as objects are social constructs. I will defend this account, in the spirit of the epistemological analysis in Chapter 8, as providing a strong enough account of the existence of arithmetical objects to complement the epistemological account of arithmetic presented in this book.

I.5 Summary

In this introductory chapter, I distinguished between two a posteriori views with respect to mathematical knowledge. According to the *epistemological* a posteriori position, mathematical knowledge is acquired empirically. According to the *methodological* a posteriori approach, empirical research is important for understanding what mathematical knowledge is like. I emphasised the need for the latter in the epistemology of arithmetic, while also accepting the importance of a priori methodology. However, empirical researchers and philosophers of mathematics do not share a common conceptual framework, which makes successful interdisciplinary research difficult. After pointing out some of the key problems, I provided a coherent conceptual framework and consistent terminology.

PART I

Ontogeny

CHAPTER 1

Proto-arithmetical Abilities

In the Introduction, I stressed the importance of distinguishing between proto-arithmetical and proper arithmetical abilities. I suggested that the term 'number' should be reserved for developed mathematical abilities dealing with exact number concepts. In particular, talk of natural numbers should be restricted to the exact number concepts associated with arithmetic. Any quantitative ability that is not properly arithmetical or mathematical should be described in terms of treating *numerosities*. In Section 1.3, I will make the distinction between proto-arithmetical and arithmetical fully explicit. But, in order to do that, we first need to have a good understanding of what kind of abilities comprise proto-arithmetical cognition.

1.1 Object Tracking System and Subitising

Recall the experiment reported by Wynn (1992), in which five-month-old infants reacted to 'unnatural' numerosity of the dolls by a longer looking time (Figure I.1). This experiment has been replicated many times in different variations and it is strongly established that it is indeed the numerosity of the objects that the infants react to, instead of other variables such as total visible surface area (for a review study of the replications see Christodoulou et al., 2017). The methodology of Wynn's experiment is also widely accepted. Infants have a limited range of reacting, so longer looking times are generally interpreted as a reaction to them having seen something surprising (Cantrell & Smith, 2013). And the infants indeed did show significantly longer looking times in the trials in which one doll and one doll put behind the screen revealed only one doll in the end. They were apparently surprised by the 'unnatural' numerosity.[1] Indeed, one replication of the original experiment showed that the

[1] By 'unnatural' I mean nothing more than behaviour of objects that would generally be considered to conflict with physical laws concerning macro-level objects.

infants were more surprised by the unnatural numerosity than they were by the actual dolls changing (from the *Sesame Street* character Elmo to Ernie, or vice versa) (Simon et al., 1995). The unnatural quantity of objects seemed to be more surprising than the unnatural changes in the *identity* of the objects. This suggests that not only is numerosity one factor in how infant observations are processed in their minds, but it is a particularly important factor. But the question is, were the infants really doing arithmetic – as claimed by Wynn – and if not, what was the cognitive capacity they were applying?

The first experiment to establish an infant ability with numerosities was reported by Starkey and Cooper (1980). They used the standard method of *habituation* to test whether infants (twenty-two weeks old) are sensitive to changes in numerosity. The infants were habituated to an array of two or three dots and then shown an array where the quantity of the dots had changed (either from three to two or from two to three). The children reacted by longer looking times. The experiment was controlled for other variables, such as the spatial arrangement of the dots, and Starkey and Cooper concluded that it was the numerosity that the infants were sensitive to. Importantly, they did not show the same ability to discriminate between the numerosity of the dots when it was increased to four and six. These results prompted Starkey and Cooper to conclude that the infants were *subitising* when discriminating between the numerosity of the dots.

The term 'subitising' was coined by Kaufman and colleagues (1949) and it refers to making fast, accurate judgements about the numerosity of observed items. The existence of this ability has been confirmed in many experiments and in adult humans it gives an alternative method to counting for determining the numerosity of objects reliably. Trick and Pylyshyn (1994) studied reaction times in determining the quantity of dots and saw a clear difference between one and four dots when compared to larger numerosities. While four and fewer items typically take 40–200 ms/item to enumerate, more than four items take 250–350 ms/item. As seen in Figure 1.1, there is a clear difference when the numerosity increases beyond four. This is because five is already beyond the *subitising range*. For numerosities from one to four, the subjects were able to subitise the answer, which is a much faster process. For numerosities larger than four, they needed to count to enumerate the objects.

What Starkey and Cooper (1980) established was that infants can already subitise. This ability seems to be behind the infant behaviour also in the experiment of Wynn (1992). Infants lack the resources to count, as

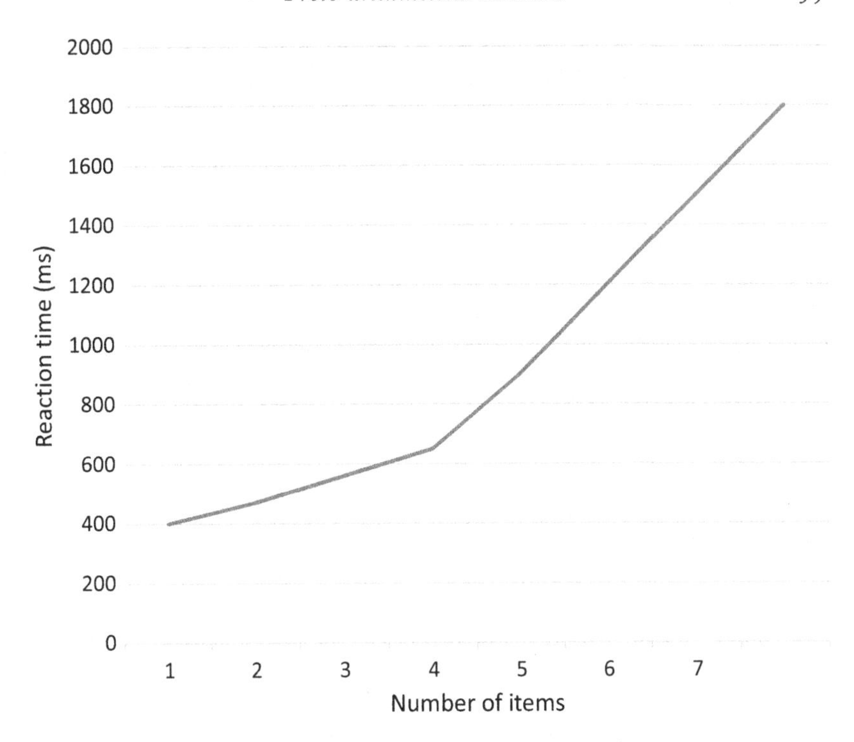

Figure 1.1 Reaction time in enumeration tasks.*
* We can see a clear difference in the reaction time (y-axis, milliseconds) when the number of dots increases over the subitising range (one to four).
Data from Trick and Pylyshyn (1994).

already suggested by their inability to discriminate between four and six dots in the experiment of Starkey and Cooper (1980), but they are able to subitise. So, what the infants could be doing in the doll experiment is subitising the numerosity of the visible dolls and then conducting calculations to create expectations concerning the numerosity of the dolls when the screen is lifted. This certainly appears to be Wynn's (1992) interpretation of the data: that subitising is a process that encodes numerical information. In short, according to her, the infants subitise the numerosity of the dolls and they possess numerical concepts that they can use to calculate expectations.

Others, however, are more sceptical. Uller et al. (1999) have suggested that a cognitively simpler explanation can explain the infant behaviour.

In contrast to the numerical concept model proposed by Wynn, they have suggested that an *object file* model can explain the same data. Object files refer to the way visual experience is processed cognitively to identify persisting objects (Kahneman et al., 1992; Noles et al., 2005). Kahneman and colleagues (1992) introduced object files as theoretical concepts to explain how successive states of objects are connected in visual experiences. Noles and colleagues (2005) have proposed that these object files can explain how our everyday perceptual experiences are formed in terms of persisting objects. The object files are closely connected to the ability to track multiple objects at once (Spelke, 2000; Trick & Pylyshyn, 1994). Three dots in our field of vision, for example, are not represented as some kind of observed 'threeness' but rather in three distinct object files.

Susan Carey (2009) has argued that object tracking is a *core cognitive* ability. Core cognition refers to the way human cognition is thought to begin with 'highly structured innate mechanisms designed to build representations with specific content' (Carey, 2009, p. 67). In this core cognitive framework, the *object tracking system* (OTS) is then considered to be the cognitive system responsible for the ability to subitise (Knops, 2020). Each observed object occupies an object file and this makes it possible to determine the numerosity of observed items without counting them. Since the object tracking ability is closely connected to the ability to individuate objects in a parallel fashion, the OTS is often also referred to as the *parallel individuation system* in the literature (e.g., Carey, 2009; Cheung & Le Corre, 2018; Hyde, 2011).

In Chapter 2, I will present in detail Carey's account of how natural number concepts are acquired on the basis of the OTS, but for now the important point is that the object tracking system and the associated object files provide an alternative explanation for the infant behaviour in Wynn's (1992) experiment. What if the dolls simply occupied object files in the minds of the infants? They expected the dolls to be persisting objects and, when there was only one doll instead of two, this violated their object file occupancy. As I see it, this explanation can accommodate the Wynn experiment in a satisfactory manner. Importantly, there is no need to presuppose that the infants were engaging in arithmetical calculations, nor that they process true numerical concepts. They could simply expect to see the dolls as persisting objects.

But what about the change from Elmo to Ernie (or vice versa), as reported by Simon and colleagues (1995)? If the above suggestion about object files is along the right lines, should the children not have been more surprised by the changing identity of the dolls than the changing numerosity of the dolls?

I believe that there is a good explanation also for this. The object tracking system, and hence the object files, are thought to be cognitively between low-level sensory processing and higher levels of cognition (Noles et al., 2005). The object files are not the kind of cognitive apparatus that is used for identifying dolls. They are much cruder, with the (putative) purpose of tracking objects across time and movement. Indeed, given the proposed character of the OTS, we would expect that infants are more sensitive to the changing numerosity than they are to the changing identity, at least as long as the latter is not more significant than a change from one doll to another.

The OTS, as should be clear by now, is not numerosity-specific. It is thought to be responsible for tracking objects and individuating them parallelly as persisting objects. However, since it makes the subitising ability possible, it is highly relevant for the topic of this book. Indeed, the OTS appears to be responsible for much of what is misleadingly called 'infant arithmetic' and 'animal arithmetic' in the literature. When infants in Wynn's (1992) experiment are supposed to do arithmetical calculations, what they could really be doing is simply applying the OTS in tracking parallel individual objects. When Hauser and colleagues observed similar behaviour in wild rhesus monkeys and Garland and Low with New Zealand robins, again the data could be explained as based on the OTS without postulating numerical concepts or arithmetical ability (Garland & Low, 2014; Hauser et al., 1996).

I do not mean to downplay the cognitive abilities of infants and non-human animals, nor to suggest that they neither observe cardinalities nor discriminate between numerosities. While the OTS is not a numerosity-specific system, it enables the subitising ability that I consider to be *proto-arithmetical*. As such, as we will see, it is highly important for explaining the development of arithmetical cognition. But it is important not to postulate cognitive abilities beyond what is needed for explaining the observed behaviour. The problem is that the ascription of arithmetical abilities can initially seem so fitting for the kind of behaviour we have been discussing. The explanation according to which infants first calculate 1 + 1 = 2, and the unfolding situation then violates their expectations, describes their behaviour in a very palpable, dare I say *exciting*, way. Yet the foundation of my methodology is that we should fight every temptation to propose such explanations if there is another, cognitively less demanding explanation available.[2] In the OTS, there is such an

[2] In comparative psychology, this principle is known as *Morgan's Canon* and it states roughly that animal activity should never be interpreted in terms of higher psychological processes if it can also be

explanation available for a lot of the observed infant and non-human animal behaviour. But not all, as we will see in Section 1.2.

1.2 Approximate Numerosity System

In Section 1.1, I have suggested that the infant behaviour in the experiment reported by Wynn (1992) and its replications can be explained by the activation of the object tracking system. However, McCrink and Wynn (2004) designed a computerised version of the experiment in which, instead of one and two dolls, five or ten blobs showed on the screen. In their study, nine-month-old infants showed similar longer looking times in cases of unnatural numerosity (equivalent, for example, to the unnatural arithmetic of 5 + 5 = 5) as the infants did in the original Wynn (1992) experiment and its many replications. But now the OTS hypothesis fails, because five and ten are both beyond the OTS range of numerosities. Certainly, the infants were not counting the items, so what was the cognitive capacity they were employing?

This area of research goes back to the nineteenth century, when the pioneering psychologists Ernst Weber and Gustav Fechner studied the way humans perceive physical stimuli. What they established was that the perception of many types of stimuli, related, for example, to brightness, loudness and weight, follows a similar law, according to which the noticeable difference between two stimuli is related to the overall magnitudes of the stimuli (Fechner, 1860). For example, humans are able to detect the difference between weights of 100 and 120 grams, but not between 1,000 and 1,020 grams, although the absolute difference in the weights is the same. They can, however, distinguish between the weights of 1,000 grams and 1,200 grams, where the ratio is the same as in the former case (100 : 120 = 1000 : 1200). This logarithmic relationship between actual physical intensity and perceived intensity has become known in the literature as *Weber-Fechner Law* (or nowadays, more commonly, simply Weber's Law) (Knops, 2020).

What does the Weber-Fechner Law have to do with the McCrink and Wynn experiment? As it turns out, one of the stimuli that humans (and many non-human animals) perceive according to the Weber-Fechner Law concerns numerosity. Data show that there is an evolutionarily developed non-verbal capacity to estimate numerosities beyond the OTS range (Dehaene, 2011). Dehaene has called this cognitive capacity *number sense* but more commonly in the literature it is referred to as the *approximate*

interpreted in terms of lower processes (Morgan, 1894). While Morgan's Canon can be problematic when used as a blunt instrument, I believe that some form of this kind of parsimony principle is justified.

number system (ANS) (e.g., Spelke, 2000). Here I will refer to this cognitive system as the *approximate numerosity system* (while retaining the same acronym ANS), in order to distinguish between the pre-verbal ANS-representations and exact number concepts.[3]

In an influential experiment, Xu and Spelke (2000) tested infants (six months of age) for the perception of numerosity stimuli. The infants were habituated to arrays of either eight or sixteen dots (Figure 1.2). When habituated to eight dots, infants reacted with surprise when presented with four or sixteen dots, but not when the array had twelve dots. When habituated to sixteen dots, they reacted to arrays of eight and thirty-two dots, but not twenty-four. This suggests that six-month-olds are sensitive to difference in numerosity when the ratio is 1:2. It has been shown that this ratio gets smaller with development. Lipton and Spelke (2003) showed that the ratio is 2:3 for nine-month-olds. This development in ANS-acuity continues all the way through pre-school and school years, normally peaking around thirty years of age (Halberda et al., 2012). However, at all stages the ANS-based numerosity estimations remain approximate and follow the Weber-Fechner Law. This can be seen in the two standard signatures of ANS-based estimation (Dehaene, 2003). First is the *distance effect*, meaning that distinguishing between numerosities becomes easier as the numerical distance between them increases. Second is the *size effect*, according to which the estimations become less accurate as the numerosities become larger.

In addition to visual stimuli, the ANS also activates from other kinds of sensory stimuli. Izard and colleagues (2009) habituated neonates with a series of tones and then presented them with a visual array of items. As in experiments with purely visual stimuli, the infants looked longer when the visual array was different in numerosity from the tones. This suggests that the ANS representations are indeed about abstract numerosities, in the sense that they are not specific to any particular type of sensory stimulus.

Not everyone agrees with this. It has been argued that, instead of evoking abstract representations numerosities, the data can be explained by cognitive systems of sensory integration involving continuous magnitudes (Gebuis et al., 2016; Leibovich et al., 2017). The main reason for this criticism of the ANS is due to what Clarke and Beck (2021) call the *numerical congruency effect*, which shows that judgements of numerosity are often influenced also by the perception of non-numerical magnitudes, such as object size. An instructive example of this was presented by Henik

[3] In the literature, ANS is also used as an acronym for *analogue* number system, referring to the same capacity (see, e.g., Geary et al., 2014).

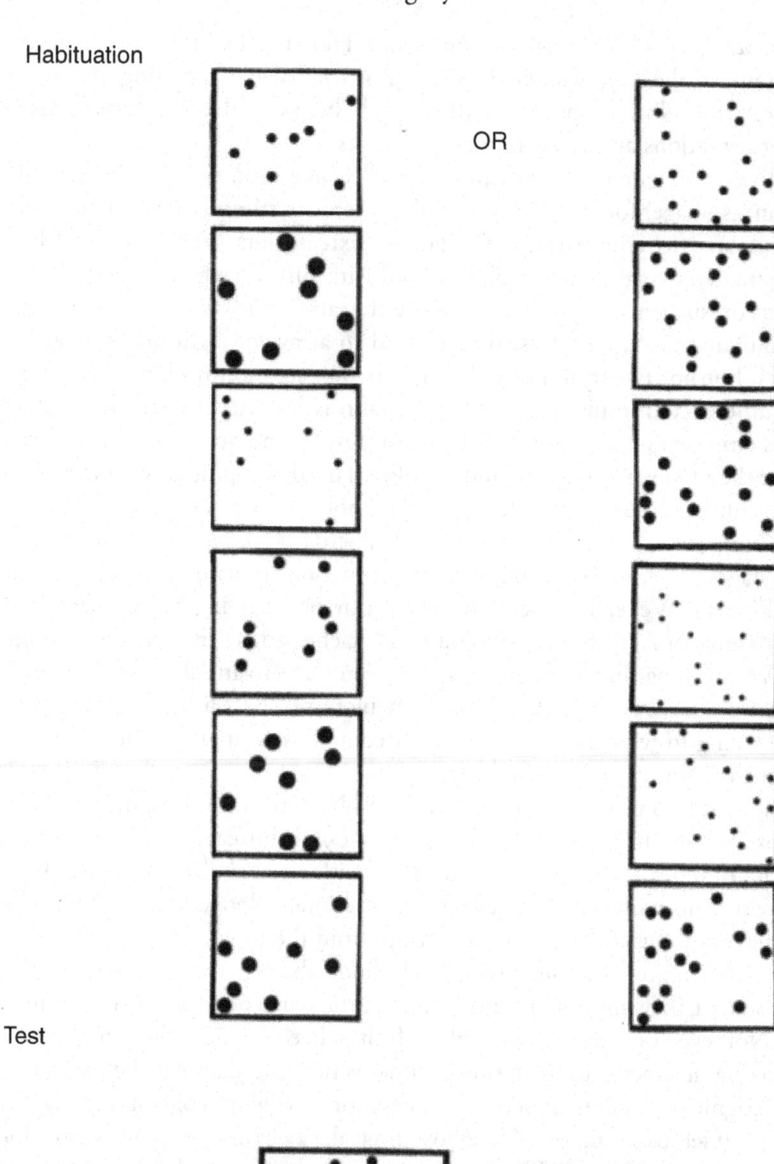

Figure 1.2 Numerosity estimation by infants.*
* Xu and Spelke (2000) habituated six-month-old infants to arrays of either eight or sixteen dots and tested their reactions to changing numerosity of dots.
Reproduced with permission.

and Tzelgov (1982), who showed that (adult) subjects show a numerical *Stroop effect*, that is, they have longer reaction times and lower accuracy in estimating which number is larger when the size of the physical numeral symbol was incongruent to the numerical value (e.g., the pair 5 3 was more difficult to process than the pair 5 3). However, as argued by Clarke and Beck (2021), this kind of argument against the ANS is problematic, since congruency effects are also present in many judgements of magnitudes. For example, judgements of duration show a congruency effect on size, length and distance (Casasanto & Boroditsky, 2008; Sarrazin et al., 2004; Xuan et al., 2007). Yet, duration is uncontroversially accepted as being perceptible and not derived from other magnitudes.

Another argument against the ANS is what Clarke and Beck (2021) call the argument from *confounds*. As pointed out by Leibovich and colleagues (2017), in the experiments with arrays of dots, for example, there are always confounding variables. When the array of dots is changed and one non-numerical variable (e.g., item size or cumulative area) is controlled for, at the same time some other variable (e.g., the cumulative area or density) also changes. How is it possible to control for all non-numerical variables and thus be certain that it is the numerosity that the subjects are sensitive to? The simple answer is that it is not. The best we can hope for is to get enough corroborating cross-modal evidence to support the hypothesis that the ANS is sensitive to numerosity. Here I cannot go into the details of the cumulating evidence, but it seems clear that in many criticisms the ANS hypothesis is put under unreasonable criteria. There are very few experiments that can be said to test sensitivity to only one property of sensory stimuli. The arguments from congruency and confounds are problematic because by the same criteria many other cognitive capacities in detaching one property (duration, size, luminance, etc.) would be in doubt. In this, the ANS does not fare any worse and, hence, I do not see a convincing reason to doubt that the ANS is indeed sensitive to numerosity.[4]

Now the question is, if the ANS represents numerosity, *how* does it do it? As we have seen, the OTS is thought to represent numerosity implicitly in object files. What is the mechanism underlying the ANS? Following the work of Dehaene (2003), it is common to speak of the ANS as representing numerosities in a *mental number line*. Again, I want to correct the terminology to *mental numerosity line*. Dehaene argued that the Weber-Fechner Law of numerosity estimations can be explained from a neural

[4] See Clarke and Beck (2021) for a more thorough discussion on the topic. However, it is important to keep in mind that they explicitly argue that the ANS is sensitive to *number*, which I disagree with.

basis by postulating an innate, logarithmic mental line on which numerosity estimations are represented (Dehaene, 2003).[5]

On a logarithmic number line, the distances between numbers become smaller as the numbers become larger. On a *linear* number line, which most students become familiar with on the primary level, numbers are evenly spaced. This question of 'logarithmic or linear' has been an important topic in research on numerical cognition. However, it is important to note that there are two different questions involved. First is whether the ANS-based *estimations* follow a logarithmic rather than linear structure. The second concerns how the ANS-based *representations* are realised in the brain. When it comes to the first question, there is little doubt (see, e.g., Nieder & Dehaene, 2009). In general, the ANS-based estimations are characterised by their logarithmic manner, following the Weber-Fechner Law: establishing the difference between, say, sets of six and seven objects is easier than that between sixteen and seventeen objects. Importantly, this is a property of the proto-arithmetical processing of numerosities. Empirical data show that, when asked to place numerosities on a line, people in cultures with limited or no proper numeral systems (such as the Pirahã and the Munduruku of the Amazon) place the numerosities in a way that is best modelled as logarithmic (Dehaene et al., 2008; Pica et al., 2004).[6]

But the second question, how the ANS-based representations are realised in the brain, is more controversial.[7] The way the ANS represents numerosities has been debated ever since Meck and Church (1983) proposed their 'accumulator' model in which numerosity information extracted from sensory input is represented on a linear scale. This account was first contested by Dehaene and Changeux (1993), who proposed a neuronal model in which the ANS-based representations are logarithmic.

[5] Interestingly, there is new research suggesting that already the pupil of the eye is sensitive to numerosity. According to a study reported by Castaldi and colleagues (2021), numerosity modulates an automatic reflex in pupil dilation, 'suggesting that numerosity is a spontaneously encoded visual feature' (p. 1). If this is confirmed, it is reason to not approach ANS in a brain-centric manner but consider it in a wider context of our neural architecture connected to sensory organs.

[6] However, this is not always the case (Núñez, 2011). Perhaps most importantly, Núñez points out that 37 per cent of the experimental runs on the Munduruku reported in Dehaene et al. (2008) showed a bimodal response that used only the endpoints of the number line. Thus, the results reported by Dehaene and colleagues are better interpreted in a conditional manner: if people place numerosities on the entire number line in non-arithmetical cultures, it is in a logarithmic rather than linear manner.

[7] Instead of representations, some authors write about 'coding schemes' (see, e.g., Dehaene, 2001b). As mentioned earlier, I am not committed to a representationalist model of the brain, so a non-representationalist is welcome to interpret this part to concern coding schemes.

This model has since received important support from new experiments, which show that the neuronal activity in the prefrontal cortex in primates is best described by a 'nonlinear compressed scaling of numerical information' (Nieder & Miller, 2003). In further research, Andreas Nieder (2016) has established that the primate brain has dedicated 'number neurons' in the intraparietal sulcus (IPS), which are thought to encode the numerosity of items in a sensory stimulus. These neurons – better called 'numerosity neurons' in the present terminology – appear to activate in association with the same numerosity independently of the sensory modality, suggesting that they represent numerosity in an abstract manner.[8] The numerosity neurons carry some 'noise', that is, when the neurons associated with the observation of seven items activate, so do some of the neurons associated with six and eight. This noise increases as the numerosities become larger, thus reflecting the Weber-Fechner Law (Nieder, 2016).[9]

If Nieder is correct, then ANS-based representations can be located in the brain, more specifically in the intraparietal sulcus. But how are the representations coded and organised neurally? A lot of further work is needed on both questions, but the data reported by Nieder suggests that the numerosity representations are in some way located in a spatially ordered fashion (Nieder, 2006). Thus, the metaphor of a mental numerosity line may not be, to some degree, only a metaphor. The well-known phenomenon of *spatial-numerical association of response codes* (SNARC) seems to support this. The SNARC effect shows that Western subjects standardly respond more quickly to small numerosities if they are on the left extrapersonal hemiside of their perceived environment, whereas they respond faster to larger numerosities if they are on the right extrapersonal hemiside of their perceived environment. This suggests that there is a spatial association of small numerosities being on the left side and

[8] Not everybody agrees. Vandervert, for example, argues that given the association of the prefrontal and parietal brain regions also with movement in space and time, it 'might therefore be more appropriate to suggest that the many research findings leading to Nieder's "number neurons" are simply about differential magnitudes related to the dynamics and kinematics of cognition and movement associated with the perceptual–cognitive control of body movement that occurred in humans, for example during stone-tool evolution' (Vandervert, 2021, p. 125).

[9] In addition to primate studies, brain imaging on humans show activation in the IPS connected to numerical tasks. Further evidence for the location of non-verbal numerosity representations in the brain comes from malfunctioning proto-arithmetical abilities in subjects who have suffered brain injury localised in the left parietal lobe, including the left intraparietal sulcus (Ashkenazi et al., 2008; Lemer et al., 2003). For more details on this topic, as well as a good general exposition of the research on non-verbal ability with numerosities, I refer the reader to Nieder (2019).

larger on the right side, as with the standard number line (Dehaene et al., 1993).[10]

The SNARC effect has also been reported in non-human animals, even in new-born chicks that, like humans, seemed to place smaller numerosities on the left side and larger ones on the right (Rugani et al., 2015). This has been seen as evidence for the early evolutionary origins of the mental number line (see, e.g., Knops, 2020, p. 61). However, Beran and colleagues (2008) failed to observe any space-numerosity association effect in rhesus and capuchin monkeys. It could well be that there is a missing factor in the experiment on chicks that made them associate smaller numerosities with the left side, and a mental number line was not responsible for this behaviour. Thus, with the present evidence, we should be careful about making inferences from the SNARC effect.

Finally, it should be noted that not all researchers agree that the ANS and the OTS are separate core cognitive systems. While this is the view held by most researchers (e.g., Agrillo, 2015; Carey, 2009; Dehaene, 2011; Feigenson et al., 2004; Hyde, 2011; Spelke, 2011), recently Cheyette and Piantadosi (2020) have proposed that there could in fact only be one innate numerosity system. One common argument for the existence of two systems is the discontinuity between identifying items in collections in the OTS range and those beyond that range. While collections of up to four items are identified almost perfectly, following the Weber-Fechner Law larger collections are identified increasingly inaccurately. Cheyette and Piantadosi, however, present a mathematical model which suggests that this difference between small and large numerosities would be part of an optimal representation of cardinal numerosities also in a single cognitive system. This question requires further study, but it is by no means certain that the OTS and ANS are indeed separate cognitive systems.[11]

The question of whether there are one or two core cognitive systems sensitive to numerosity of course prompts the question of just what a core cognitive system *is*. Every cognitive system comparable to the OTS and the ANS is a higher-level abstraction of neuronal activity. As such, we should not be ontologically committed to cognitive systems beyond what is

[10] The direction of SNARC effect does not vary with handedness, but it does change with the direction of writing. Iranian and Arabic subjects accustomed to right-to-left writing responded faster to larger numerosities on the left extrapersonal hemiside (Dehaene et al., 1993; Zebian, 2005). Thus, the evidence points to spatial association with numerosity size being universal, but the direction depending on the cultural custom of writing.

[11] In Pantsar (2014), I supported the one-system theory myself, but in the face of the state of the art in empirical research, in later works I have focused on the two-system model.

necessary for explanations. If Cheyette and Piantadosi are correct and there can be a mathematical model that necessitates only one system, with an Occam-like principle it would seem to be preferable over two-system models. However, a mathematical model should also be consistent with brain-imaging data. Supporting the two-system model, subitising and estimating have been reported to have different neural correlates in an fNIRS (functional near-infrared spectroscopy) study (Cutini et al., 2014).[12]

In this book, I will follow the two-system model, but I recognise that this remains an open question. As of now, the evidence still appears to favour the existence of the OTS and the ANS as distinct systems, and the association of the subitising ability with the former and the estimation ability with the latter. It could be that this understanding of the core cognitive abilities needs to be adjusted, but for the present approach that would not make a dramatic difference. The account I will be developing is compatible also with a single-system account. Instead of the particular core cognitive systems, the present account makes important use of the subitising and estimating *abilities*. That these are two different abilities is much less controversial than the two cognitive systems being different.[13]

1.3 Proto-arithmetic and Arithmetic

In Sections 1.1 and 1.2 we have seen that many experiments with infants and non-human animals can be explained by them employing the object tracking system or the approximate numerosity system (or perhaps both), with no reason to postulate more sophisticated numerical abilities to explain the behaviour. However, not all experimental data can be explained this way. Cantlon and Brannon (2007) conducted an experiment in which macaques were shown two configurations of dots on a screen (for 500 milliseconds each, corresponding to, for example, $1 + 1 = 2$ or $2 + 2 = 4$) and then were made to choose between two options on a

[12] Recently there have also emerged results from AI research that have been interpreted to question the existence of a numerosity-specific system such as ANS. These results show that generic artificial neural networks going through unsupervised learning develop a human-like acuity in numerosity estimation, suggesting that, instead of ANS being numerosity-specific, it could be a more general characteristic of learning through visual observations (Stoianov & Zorzi, 2012; Testolin et al., 2020). While these results are very early and tentative, they could point to an interesting new research direction in the development of numerical cognition. For more on this topic, see Pantsar (2023a).

[13] This state of affairs was very different at the turn of the century. Dehaene (1997), in the first edition of *The Number Sense*, for example, was a proponent of the view that the subitising ability is fundamentally a part of the estimation ability. See Section 2.3 for more.

screen for the correct sum. What they observed was that the macaque performance was largely comparable to that of college students in the same non-verbal task. The conclusion of Cantlon and Brannon was that the 'data demonstrate that nonverbal arithmetic is not unique to humans but is instead part of an evolutionarily primitive system for mathematical thinking shared by monkeys' (Cantlon & Brannon, 2007, p. 1).

These data are difficult to explain by simple mental numerosity representations of cardinality. What the macaque were reported to be doing seem to be genuine *operations* on numerosities. What this suggests is that the OTS- and ANS-based numerosity representations can be manipulated mentally. This is an important finding and is corroborated by many experiments. Indeed, the macaque behaviour is far from the only reported infant and non-human animal behaviour concerning numerosities that is truly remarkable (see, e.g., Agrillo, 2015; and Pepperberg, 2012 for studies on new-born chicks and parrots, respectively). Furthermore, it seems that numerosity is the decisive factor in most of the reported cases. Yet, again, we should be cautious not to postulate more sophisticated cognitive abilities than is needed to account for the behaviour. That is why, also in this case, it is important to stick to the distinction between proto-arithmetic and arithmetic.

Due to the short reaction times required in the experiment reported by Cantlon and Brannon, the college students could not revert to their arithmetical ability. With more time, they could have easily counted the dots and outperform the macaques. One main point of the experiment was to deny them that chance, in order to test their 'nonverbal arithmetic' skills against those of the macaques. However, while it is possible that there are nonverbal arithmetical skills, the experiment of Cantlon and Brannon did not test them. What it tested was the way macaques and humans use their proto-arithmetical, evolutionarily developed, skills in cardinality processing tasks. As such, however, the results are highly important. They show that monkeys can process operational transformations of numerosities. Indeed, another study by Cantlon and colleagues showed that this monkey proto-arithmetical ability carries some of the same signatures as human (proper) arithmetical ability, like the problem size effect, according to which reaction times and error rates increase as the numerosity sizes become larger (Cantlon et al., 2016).

Yet even this should not make us think of human arithmetical ability merely as a specification of evolutionarily developed proto-arithmetical abilities. The problem size effect, for example, is a common characteristic of both proto-arithmetic and arithmetic, but this can be due to different factors. While with the ANS the problem size effect is due to the

Weber-Fechner Law, with mental arithmetic it can be due to the increasing computational and cognitive *complexity* of the task (Buijsman & Pantsar, 2020; Pantsar, 2019a, 2021b). Due to limitations in working memory and attention span, for most people arithmetical calculations quickly become prohibitively complex to be carried out mentally (see, e.g., Imbo et al., 2007). It is possible that there are some common grounds for the problem size effect for proto-arithmetical tasks and arithmetical tasks. But it is also likely that, due to the important differences in the tasks – for example, only the latter including verbal or symbolic processing – the two effects have partly different causes. For this reason, we cannot assume that arithmetical calculations are a developmental continuation of proto-arithmetical transformations of numerosities.

Nevertheless, there are important connections between arithmetical and proto-arithmetical processes. For example, consider the following extremely simple task: Which one of the numbers below is bigger?

4 5

Compare that to the task of finding out which of these numbers are bigger:

4 9

One would not expect arithmetically skilled humans to show differences in reaction times, but the data show that the top pairing takes considerably longer than the bottom pairing. This is consistent with the distance effect: the larger the numerical distance between two numbers, the faster we are in solving the problem (Dehaene, 2011, pp. 62–64). It is interesting that the distance effect remains even when the test subjects are trained to solve the problem. It even remains when the quickest way of solving the problem only requires comparing the first digits of multi-digit numbers. For example, consider the pairing:

71 65

Now compare it to the pairing:

79 65

In this case, one would certainly expect the reaction times to be similar, given that in both cases it only needs to be established that 7 is bigger than 6. Yet the distance effect remains (Hinrichs et al., 1981; Pinel et al., 2001). It even remains when the task is to determine whether two numbers are the *same*, in which case it suffices to only determine that two physical shapes are different, as well as when numeral words replace numeral symbols (Dehaene & Akhavein, 1995).

The generally accepted explanation for all these results is that seeing numeral symbols (or numeral words) automatically activates the ANS and makes us process the numerosities as magnitudes (Dehaene, 2011, p. 16). Conforming to the Weber-Fechner Law, when the numerical distance is larger, the task of choosing the bigger number gets easier. It is important to note that the distance effect, unlike the problem size effect, is not a characteristic of arithmetical cognition. But even in tasks where arithmetical skills would be all that is needed, and the proto-arithmetical skills indeed only delay the correct solution, it seems that we cannot escape our proto-arithmetical origins. Even when trained for particular tasks that do not require assessing the numerical size, subjects cannot help processing it.

What these, and many other, experimental data suggest is that there is a connection between proto-arithmetical abilities and arithmetical abilities that endures also in arithmetically skilled individuals. Thus, the question becomes how the proto-arithmetical abilities are employed in the development of arithmetical abilities and arithmetical knowledge. In early literature on the development of numerical cognition, it was thought that arithmetic is made possible by first forming semantic representations of numerals and numeral symbols (e.g., McCloskey, 1992; McCloskey & Macaruso, 1995). This was contrasted by the 'triple-code' model of Dehaene and Cohen (1995), in which there are thought to be different representations of numerosities (visual symbolic, auditory verbal and approximate numerosity estimations). In the triple-code model, these representations have different roles in numerical cognition tasks. This model has been further refined by Campbell and colleagues (Campbell, 1994; Campbell & Epp, 2004), whose 'encoding complex' model includes interactions between the different kinds of representations.[14] In Chapter 2, we will focus on the issue of how the representations in the triple-code model are made possible. Indeed, so far with regard to the ANS, we have only focused on the third type of representation, that of the approximate numerosity estimations. Thus, the big question is how these proto-arithmetical representations – as well as those provided by the

[14] That there are different representations is supported by data showing activation in different brain areas (Nieder, 2019, p. 211). In a particularly interesting finding, it has been established that seeing numeral symbols is associated with activation of a special area of the temporal lobe, the ventral occipito-temporal cortex, also called the 'number form area' in the literature (Shum et al., 2013). This area shows increased activity with numeral symbols compared to other symbols, which can also help explain why dyslexia and *dyscalculia* (difficulty in learning arithmetic) are distinct learning disabilities (Nieder, 2019, p. 188).

OTS – are employed in acquiring exact number concepts, that is, arithmetical representations of cardinalities.

Before we move on to that question, it should be noted that not everyone wants to make the present distinctions between proto-arithmetic and arithmetic, as well as numerosities and numbers, even when aware of the kind of problems I have been discussing. I assume that, by now, most empirical researchers have encountered the kind of criticism that I have presented against incongruent use of terminology, yet they do not seem to have a problem with continuing to write about 'infant arithmetic' or 'animal arithmetic'. For the most part, such issues do not seem to be seen as particularly important among empirical researchers. Some authors, however, explicitly justify their choice of terminology. Carey, for example, writes:

> In the literature on mathematical cognition, analogue magnitude number representations are sometimes called 'numerosity' representations, for they are representations of the cardinal values of sets of individuals, rather than fully abstract number representations. There is no evidence that animals or babies entertain thoughts about 7 (even approximately 7) in the absence of a set of entities they are attending to. Still, cardinal values of sets are numbers, which is why I speak of analogue magnitude number representations rather than numerosity representations. (Carey, 2009, p. 136)

I can understand the practical side of using the more familiar word 'number' instead of the technical term 'numerosity'. But Carey's reason for her use of vocabulary is hardly satisfactory. While it is correct that (cardinal) numbers are measures of cardinality of sets, so are (cardinal) numerosities in the sense used by myself and others (e.g., De Cruz et al., 2010; Dos Santos, 2022). The entire point of introducing a new technical term such as 'numerosity' is to differentiate between two ways of representing the cardinal values of sets. Hence, Carey's reasoning for sticking with the term 'number' is quite problematic.

Another explicit justification for speaking about numbers instead of numerosities in the context of the ANS has recently been given by Clarke and Beck (2021). They argue that:

> [T]he ANS represents numbers (i.e., that numbers serve as the referents of the ANS), but under a unique mode of presentation that respects the imprecision inherent in the ANS . . . This will allow us to avoid a commitment to exotic entities such as 'numerosities' without losing sight of the important differences between ANS representations and the precise numerical concepts that emerge later in development. (Clarke & Beck, 2021, p. 5)

Like others – for example, Dutilh Novaes and dos Santos (2021) and Núñez et al. (2021) – I find this justification equally unsatisfactory to Carey's. In my terminology, the ANS represents *cardinality* in terms of *numerosity representations*. With these distinctions in place already in the terminology, it is impossible to lose sight of the differences between the ANS representations and precise numerical concepts. Knowing how often in the literature problems have arisen because the two have been confused, why not distinguish between them in the terminology? If the reason is simply that numerosities as entities are considered to be 'exotic', surely that is a weak justification. As Clarke and Beck concede, we need to distinguish between ANS representations and exact number concepts in any case. So, if there is an exotic entity involved, it is already included in the ANS representations. It only makes sense that we would use a different scientific term when introducing theoretically novel entities like that. Hence, to conclude this chapter, I believe it is more important than ever to distinguish between proto-arithmetic and arithmetic, as well as between numerosities and numbers. In Chapter 2, we will fully grasp why, as we tackle the question of how number concepts can arise on the basis of numerosity representations.

1.4 Summary

In this chapter, I presented the *proto-arithmetical* abilities, that is, *subitising* and *estimating*, and emphasised the importance of distinguishing them from arithmetical abilities. I reviewed the empirical literature on the cognitive basis of proto-arithmetical abilities, focusing on the *core cognitive* theory of the object tracking system (OTS) and approximate numerosity system (ANS). Although the topic requires further research, I proceed with the view that the OTS and ANS are different cognitive systems and responsible for subitising and estimating abilities, respectively.

CHAPTER 2

Acquisition of Number Concepts

In Chapter 1, we have gained a better understanding of what is involved in the supposed infant and non-human animal numerical and arithmetical abilities reported in the empirical literature. Indeed, what we established is that such abilities should not be considered to be numerical or arithmetical at all but are more accurately described as proto-arithmetical. However, as my choice of the term 'proto-arithmetical' already suggests, there are very good reasons to think that proper arithmetical ability builds on these evolutionarily developed abilities. What needs to be explained now is how this development takes place in human ontogeny. In most modern cultures, basic arithmetical abilities are learnt by the great majority of children before the age of ten. What happens in those few years during which children turn from only possessing proto-arithmetical abilities to being (more or less) proficient practitioners of arithmetical calculations? In this chapter, I focus on the first – or at least one of the first – stages in this development, that of acquiring stable and exact number concepts. I end up endorsing a form of *bootstrapping* account of number concept acquisition, originally presented by Carey (2009) and further developed by Beck (2017). However, unlike them, in my version of the bootstrapping account I see an important role for the ANS (in addition to the OTS). In addition, in Chapter 3 I will update the bootstrapping account to include a specific notion of cultural influences, namely the *enculturation* account presented by Menary (2015).

2.1 Pre-conceptual and Conceptual Abilities with Numerosities

In order to be able to do proper arithmetic, children first need to acquire number concepts. Unlike ANS-based representations, these concepts need to concern exact, discrete and linear cardinality representations. Unlike the OTS-based representations, they cannot be limited to small numerosities. But when can a child be said to have acquired number concepts? And what precisely should count as acquiring number concepts? Natural numbers, of course, form an infinite collection, and our finite brains cannot contain an infinite amount of number concepts. So, should we focus on the question when children have acquired a *general* notion of natural number? But what does that mean? If it refers to the kind of exact mathematical notion as natural numbers as fixed by, for example, the Dedekind–Peano axioms (Peano, 1889), it would be a notion that not even most adults would possess.[1] This seems to be an unreasonably strict criterion, given that most adults (and children of school age) possess considerable arithmetical knowledge and skills.

For this reason, it makes sense to look for weaker criteria for number concept acquisition. In the case of natural numbers, such criteria often concern the ability to *count* with number concepts. However, as noted by Benacerraf (1965), the word 'counting' refers to two different types of processes. In *intransitive* counting, one simply recites the numeral word list in the right order, like in the beginning of a game of hide and seek. In *transitive* counting, items on the numeral word list are used to enumerate items in collections, like when counting apples on a table. Transitive counting is often also called *enumeration* in the literature, in which case the term 'counting' usually refers to intransitive counting (see, e.g., Rips et al., 2008).

It has been established that children are able to do intransitive counting much before they master transitive counting, but also that transitive counting is not enough to possess number concepts (see, e.g., Davidson et al., 2012). Following Wynn (1990), one standard method of ascertaining children's development in numerical cognition is the so-called give-*n* test. In this test, a child is presented with a collection of objects and asked to give *n* of them. If the child consistently gives exactly *n* objects, they are thought to grasp the number concept *N*. However, as Davidson et al. (2012) point out, there is a stage in development when children can do transitive counting up to *n* but fail the give-*n* test. Thus, even the ability to do transitive counting is not sufficient for possessing number concepts.

When children begin to pass the give-*n* test, they learn the first numbers in ascending order. Usually at roughly 2 years of age, they pass the give-1 task, becoming *one-knowers*. Then, in stages which typically take four to five months each, children become *two-knowers*, *three-knowers* and

[1] The Dedekind–Peano axioms (often simply called Peano axioms) are nine simple formal rules that define natural numbers and their arithmetic. They can be found on p. 133.

four-knowers (Knops, 2020).[2] At that point, children grasp something more general about cardinal numbers and in addition to the give-5 test, they start passing the give-*n* test also for six items, seven items and so on (Lee & Sarnecka, 2010). When children are in this way able to more generally match the last numeral word uttered in the counting sequence with the cardinality of a group of objects, they are said to have become *cardinality-principle* (CP)-knowers (Lee & Sarnecka, 2011; Sarnecka & Carey, 2008).

Clearly children need to be cardinality-principle-knowers in order to possess proper exact number concepts, but for that, more than being able to solve the give-*n* task is required. Piaget (1965) reported a famous experiment which suggests one reason why. He had children count two rows of marbles containing an equal amount of items. After the children had agreed that there were an equal amount of marbles, he spread the marbles on one row further apart to cover a longer distance. When asked which row has more marbles, children before five years of age typically answered that the spread-out row did. Piaget's own conclusion was that children before that age do not yet possess abilities with numerosities, namely that children do not grasp numerosity conservation. We now know that this is mistaken, since already infants have a sense of numerosity conservation. But Piaget's result can still be important in other ways because it suggests that children, in this experiment at least on the *four-knower* level, did not yet properly grasp number concepts.[3] This prompts the need to detail just what is involved in possessing number concepts and the ability to count.[4]

[2] It should be noted that these are typical timeframes for the development, but children show significant variation in all kinds of cognitive development related to number concepts.

[3] Dehaene (2020, pp. 163–164) suggests another explanation: children do possess number concepts but they confuse two features of the stimulus: physical size and numerosity. This is related to the Stroop effect in adults, which refers to difficulty in separating incongruent stimuli, like in the case of determining which of the numbers 5 3 is greater (Henik & Tzelgov, 1982). While there may be similarity, I am not convinced that the behaviour Piaget observed is comparable to the Stroop effect, or other similar phenomena reported in the literature. As reported by Piaget, young children systematically fail in the task, whereas the Stroop effect in adults shows primarily as longer reaction times. Based on current knowledge, I believe it is more likely that children who fail the Piaget task do not yet possess fully formed number concepts.

[4] Piaget's result has been challenged also from other directions. In perhaps the most interesting one, McGarrigle and Donaldson (1974) presented the new arrangement as done by a malicious teddy bear. Now the children answered correctly that both rows have equally many marbles. So, what the children may have done in the original experiment is to want to please the adult experimenter: if the adults ask which row has more after the rearrangement, something must have changed. These types of results have led to new ways of running experiments in which 'talking' puppets are used as proxies to avoid the adult pleasing effect.

Gelman and Gallistel (1978) proposed a pioneering account. According to them, there are five counting principles that a child must master. In the present terminology, this kind of counting goes beyond intransitive and transitive counting. It is counting with *number concepts*, that is, counting so that the result of the counting process is grasped in terms of number concepts. Paraphrased for the approach here, the principles are as follows:

1. One-to-one: Every item in the collection being counted is paired with exactly one numeral word.
2. Stable order: The way in which numeral words are used to tag items must follow a stable order.
3. Cardinality: The numeral word used for tagging the last item in the collection expresses the cardinality of the collection.
4. Abstraction: Counting procedures are not tied to any particular items, types of items or arrangement of items.
5. Order irrelevance: Counting procedures to not depend on the order in which the items are tagged.

Criterion (3) is the cardinality principle and criteria (1) and (2) clearly must be fulfilled in order for the child to be a CP-knower (otherwise they would fail to establish the cardinality of the items due to a flawed use of numeral words in the tagging procedure). But now we can see a plausible reason why the children became confused in Piaget's experiment. They lacked a proper grasp of *abstraction* in the counting procedure, which in Piaget's (1970) theory only emerges in later stages of cognitive development.

Grasping abstraction indeed seems to be a key step in number concept acquisition. Only by grasping the abstraction principle can the child understand how the counting procedure is related to determining the cardinality of a collection. In Piaget's experiment, the children were clearly quite close to grasping this relation. After all, they were able to establish the cardinality of the collection of marbles in the first stage of the experiment. What they did not yet grasp was that the cardinality does not change if the arrangement is changed. They confused another method of measuring size, namely length, with that of cardinality. Thus, a likely reason for this is that the children did not yet possess fully developed number concepts. They had enough knowledge about counting to pass the first stage, but they had not yet fully grasped the property that numbers capture, namely cardinality.

But what kind of ability with numerosities *did* the children possess in that developmental stage? Clearly, they were already considerably above the level of the proto-arithmetical abilities of subitising and estimating, but their ability was not yet arithmetical. The children did not possess fully developed natural number concepts, but did they possess *some* kind of number concepts? This is a key question in explaining the development of arithmetical knowledge in individual ontogeny. What exactly are number concepts and at what developmental stage do children typically acquire them? Which cognitive abilities, in particular proto-arithmetical ones, are employed in this process? In this chapter, we will see that these questions have received radically different answers in the literature.

2.2 Number Concept Acquisition: Nativism

Since Gelman and Gallistel did important pioneering work in the research on counting and number concept acquisition, let us start from their account. In Section 2.1 I have presented the counting criteria of Gelman and Gallistel in terms of numeral words, but it should be noted that their original account also allows for non-verbal counting (numeral words being replaced by the more general 'tags'). This is not merely a terminological difference, since the formulation of Gelman and Gallistel is closely tied to their understanding of number concept acquisition, which is based on the idea that number concepts are *innate* (Gallistel & Gelman, 1992). Such *nativist* positions are relatively common in the literature, but they take many different forms. Gelman and Gallistel hold number concepts to be pre-verbal and innate (Gallistel, 2017; Gelman & Gallistel, 2004). Wynn believes infants to have arithmetical abilities and exact number concepts (Wynn, 1992). These are clearly nativist positions. However, with other researchers it is not always clear whether their account should count as nativism. Butterworth (1999), for example, has argued for an innate 'number module' and, as we have seen, Dehaene (2011) argued for an innate mental number line. Nevertheless, the problem of incongruent terminology aside, I do not believe that either account should be understood as a nativist one concerning number concepts. The shared key feature between the accounts of Butterworth and Dehaene is the assumption that there is an innate capacity for observing numerosities, but this is best understood as dealing with proto-arithmetical abilities. In contrast, the shared key feature of the accounts of Gelman and Gallistel and Wynn is the assumption that number concepts are innate. For Wynn, the innateness is also extended to (basic) arithmetical abilities, a view also supported in De Cruz and De Smedt (2010).

I do not agree with any of the nativist accounts when it comes to the kind of theoretical assumptions they make about innate numerical or arithmetical capacities. The source of this disagreement is more than just the incongruent terminology. This is not to downplay the terminological confusions in the literature. In terms of terminology, Dehaene's account, for example, could be seen as a nativist one, given that he writes about infant knowledge about arithmetic. However, his theory can – and, as I interpret it, should – be understood in terms of numerosity representations involved in proto-arithmetical abilities being innate, while retaining a non-nativist stance on number concepts. I fully agree with this position. Indeed, with the terminological confusion clarified, Dehaene's account reflects my main reason for being sceptical about all the proposed nativist accounts. All of them get their putative empirical support, in one way or another, from the data on infant and non-human animal cognition. But, as we have seen, those data exclusively concern proto-arithmetical abilities. To the best of my knowledge, there is not a single piece of data that warrants the postulation of proper arithmetical abilities for human infants or non-human animals. Likewise, I am not aware of any data that implies the existence of proper number concepts in human infants or non-human animals.

Instead, there are data that specifically support the position that number concepts are *not* innate. I assume that, if number concepts were innate, they would be universal evolutionary adaptations, just like the capacity for proto-arithmetical abilities is. However, researchers agree widely that, in cultures like the Pirahã and the Munduruku, people do not possess number concepts, even though they possess similar proto-arithmetical capacities as people in other cultures (Everett, 2017; Gordon, 2004; Pica et al., 2004). Of course, it is possible that proto-arithmetical capacities and number concepts have emerged at different stages of biological evolution, and that can explain why the Pirahã and the Munduruku possess the former but not the latter. This, however, is unlikely given that both peoples have branched out relatively recently from people who do possess number concepts (Pantsar, 2019b). In addition, bi-lingual Munduruku children who also speak Portuguese learn to count and acquire number concepts (Everett, 2017; Gelman & Butterworth, 2005). These findings suggest strongly that, instead of lacking number concepts because of different genetic material, the Pirahã and the Munduruku do not possess number concepts for other reasons, presumably connected to the fact that their languages do not contain numeral words (Gordon, 2004; Pica et al., 2004).[5]

[5] This argument will be developed in detail in Chapter 4.

It should be noted that this argumentation is not enough to reject the nativist accounts. It is not necessary that every innate capacity is realised in all individuals or cultures. In this way, the innateness hypothesis could still be consistent with the existence of anumeric cultures, such as the Pirahã and the Munduruku. Nevertheless, I believe that the burden of proof lies now on the nativist side: given the existence of many anumeric cultures, there would need to be some evidence supporting the existence of innate number concepts. Importantly, this evidence should not be explainable based on proto-arithmetical abilities. For this reason, in order to make sense of the literature, we need to be clear what the supposed innate concepts are. To give an example, Leslie and colleagues write that: '[T]here is an innate system of arithmetic reasoning with preverbal symbols for both discrete and continuous quantity. On this view, the preverbal system supports arithmetic operations from the outset and directs the learning of the number words with the strict ordering that is a prerequisite for valid counting' (Leslie et al., 2008, p. 213). Taken literally, Leslie and colleagues suggest that symbolic arithmetic is an innate capacity. However, what they are actually referring to are the pre-verbal proto-arithmetical abilities of subitising (discrete) and estimating (continuous), and how they direct the learning of symbolic arithmetic. Thus, we can see that a seemingly nativist position over number concepts is in fact something much weaker, a nativist position over proto-arithmetical capacities, which is not controversial in any sense.

This is not an isolated incident. I suggest that, with the proper distinction between proto-arithmetical and arithmetical abilities in place, the supposed evidence for nativism is better understood as evidence for nativism concerning proto-arithmetical abilities. Number concepts, on the other hand, are acquired later in ontogeny and seem to require specific cultural conditions. Hence, we need to consider accounts of number concept acquisition from a new perspective. It should be noted, however, that we cannot expect an alternative account to *demonstrate* the non-existence of innate number concepts. I believe that, based on research on proto-arithmetical abilities, we can dismiss the possibility that human infants possess number concepts and arithmetical abilities. We cannot, however, rule out the possibility that number concepts are innate, but they take a long time to emerge in children and are not universally instantiated. However, currently there is no evidence or a positive argument why we should make such an assumption. Indeed, as we will see in Section 2.4, there are non-nativist accounts that can explain number concept acquisition while not making any problematic assumptions. In combination with the lack of support for nativism, this should be considered to be indirect evidence against the innateness of number concepts.

2.3 Number Concept Acquisition: Number Sense

In early accounts of how children learn transitive counting, the ANS was generally seen as the key proto-arithmetical cognitive system (see, e.g., Gallistel & Gelman, 1992). Perhaps most importantly, in the first 1997 edition of his highly influential book, *The Number Sense*, Dehaene argued that the estimation ability due to the approximate numerosity system – or in his terminology, the number sense – is the sole proto-arithmetical ability that is needed to explain the data on infant and non-human abilities with numerosities. For collections of one to three items, Dehaene's idea was that, rather than a separate subitising capacity, the cognitive process is a form of 'precise approximation' (Dehaene, 1997). However, in the second (2011) edition, he admitted to being wrong and accepted that further experiments (e.g., Revkin et al., 2008) establish subitising firmly as its own ability, with its own functional characteristics (Dehaene, 2011, p. 257). This laudable change in Dehaene's thinking should be kept in mind when reading his earlier work on number concept acquisition. However, since much of that earlier account remains unaffected, we should be safe in addressing his general model of number concept acquisition, as long as we make the necessary adjustments.

In Section 1.2, we encountered the logarithmic neuronal model Dehaene and Changeux presented for the approximate numerosity sense (Dehaene & Changeux, 1993). The crux of the functioning of this model is that numerosity is represented neuronally in an approximate fashion in logarithmically coded *numerosity detectors* (Figure 2.1). Each neuron in this set-up fires in connection to a preferred numerosity, but this firing has a tuning curve of a Gaussian function of the logarithm of the numerosity (Dehaene, 2007). Thus, as seen in Figure 2.1, the numerosity detectors have partial overlap. Four and eight objects are easy to distinguish because the peaks of the tuning curves are largely separate. With, say, seven and eight objects the peaks are close together, explaining why the ANS is not able to consistently distinguish between them. Originally, Dehaene thought that the different behavioural signatures for one, two and three objects is due to sharper tuning curves that allowed encoding a precise value, but in the updated version he accepts that this is due to them being due to applying a different cognitive system, that is, the OTS (Dehaene, 2011).

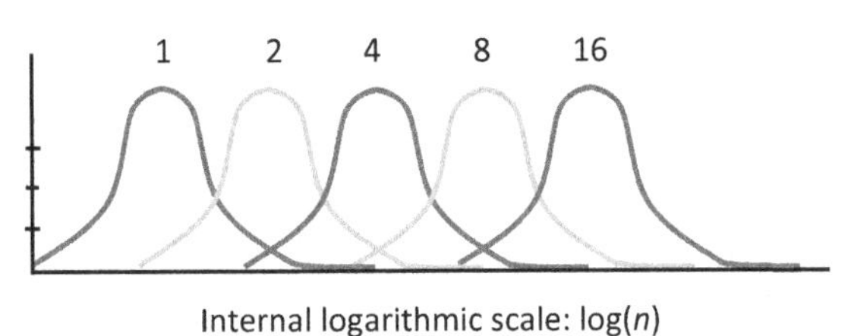

Figure 2.1 The Dehaene–Changeux neuronal model of numerosity detectors. After Dehaene (2007).

Keeping in mind that the first three (or four) number concepts may emerge differently due to their relation to the OTS, how do number concepts generally emerge in Dehaene's model? His overarching idea was that, as the tuning curves get sharper, our numerosity representations become more precise. In Dehaene (2007) he identified the acquisition of numeral symbols as a transformative stage in this development. Indeed, he saw two ways in which numeral symbol acquisition changes the innate 'mental number line'. First, numeral symbols enable sharpening of the tuning curves and thus increasing precision. Second, numeral symbols change the coding scheme of the mental numerosity line from logarithmic to linear. This second change is very dramatic indeed. While the ANS is characterised by its logarithmic neural and behavioural signatures, Dehaene (2007) accepted that, through the acquisition of numeral symbols, the nature of the coding scheme is changed into a linear one. This prompts the question: does it still make sense to talk of the same numerosity representations? Or have the numeral symbols changed the representations so drastically that new representations have emerged, ones that may be connected to the ANS-based representations but should be distinguished from them?

I believe that a partial answer can be found in the second edition of *The Number Sense*, where Dehaene accepts that the OTS also plays a role in number concept acquisition:

> [The] coding principle [of subitising] is radically different from the way that numbers are encoded on the approximate mental number line. Here, numbers are represented through noisy distributions of activation, such that seven and eight overlap, while two and eight do so far less. There is nothing in the approximate number system to support a system of exact arithmetic with discrete numbers. With the object file system, however, we can track each object precisely (as long as their number does not exceed three). The concept of 'natural number', the cornerstone of our arithmetic system, probably arises from our remarkable capacity to track small numbers of

> objects, combined with our intuitive number sense, which tells us that any set, however large, has a cardinal number. Somehow, around the age of 3 or 4, these two systems snap together. Suddenly, children infer that any set must have a *precise* number, and that 13 is therefore a distinct concept, radically different from its neighbours 12 and 14. This mental revolution, unique to *Homo sapiens*, is the first step on the way to higher mathematics. (Dehaene, 2011, p. 260)

In this passage, we see that Dehaene has changed his view quite radically. He explicitly states that the ANS does not support exact arithmetic with discrete numbers. Instead, the object file system (i.e., the OTS) is thought to be the cognitive basis of the concept of natural number, while the ANS only plays a role in a later stage, when it 'snaps together' with the OTS. As we will see in the next section, these ideas prove to be very fruitful in explaining number concept acquisition.

2.4 Number Concept Acquisition: Bootstrapping

In this section, I will pursue the idea that the OTS is integral for number concept acquisition. But let us first track back a little and recall what is known about number concept acquisition in ontogenetic development. Since nativism concerning proper number concepts is unfeasible, number concepts must be something that individuals acquire in their ontogeny. In the early psychological accounts of number concepts, such as that of Piaget (1965), this assumption played an integral role. We have seen that Piaget's account was flawed because it denied that children (typically) before the age of five can have any abilities with numerosities. What Piaget thought was that all representations of numerosities arise from logical capacities, which take years to develop.

We now know that Piaget was wrong. The data discussed in Chapter 1 clearly show that children can mentally represent numerosities in early infancy. However, Piaget was right in that it takes children longer to possess proper number concepts.[6] In Section 2.1 we have seen how the empirical data suggest that number concept acquisition happens at a distinctive order and pace. Children grasp the first four number concepts in an ascending order, after which they grasp the concept of number more generally and become cardinality principle-knowers. To use the Fodorian notation, when children acquire the number concepts ONE, TWO,

[6] Although, as seen in Section 2.1, the first proper number concepts appear much before the age of five.

THREE and FOUR, it is always a relatively lengthy process.[7] However, acquiring the concepts FIVE, SIX, SEVEN, etc. is different, as children learn the concepts at roughly the same time. This phenomenon prompts the question of what happens after the concept FOUR?

Here it is important to remember that we are discussing number concepts and not numeral words. As any parent can tell, children learn (both intransitive and transitive) counting word by word at a pace that does not carry the signature of number concept acquisition. This is because, in learning the counting sequence, children do not yet possess number concepts. What children can do at this stage is recite the counting list of numeral words in the right order, but they do not yet grasp what numbers are, as evidenced by their failure in the 'give-*n*' task (Davidson et al., 2012). But, after children have become CP-knowers and acquire number concepts, this has changed. At this stage, children can consistently also associate the right numeral with numbers larger than four.

The reader may already suspect that it is not a coincidence that the last level of being a subset-knower is being a four-knower. After all, the limit of the object tracking system was four objects. Indeed, one of the most researched contemporary accounts of number concept acquisition draws a direct connection between the OTS and the process of becoming a CP-knower. This is the *bootstrapping* account originally formulated by Susan Carey (2009) and developed further by Jacob Beck (2017). As presented by Carey, the bootstrapping process consists of three stages. First is the acquisition of a counting list, in which numeral words function as place-holders without semantic content. At this stage, children can recite part of a counting list without grasping its connection to cardinalities, that is, they fail the 'give-*n*' test. The most likely explanation for this failure is that children at this first stage do not yet possess number concepts. They do grasp intransitive counting, that is, reciting the counting list of numeral words, and they may be able to tag objects in connection with reciting the word list. However, they do not properly engage in transitive counting with number concepts, since the process does not establish the cardinality of the collection of items being counted.

At the second stage of the bootstrapping process, the OTS plays a fundamental role. As described in Section 1.1, the OTS is not thought to represent numerosities explicitly. In observing four objects, there is no

[7] Recall that the all-uppercase ONE is used to refer to the mental concept possessed by an individual, to be distinguished from the numeral word 'one', the numeral symbol 1, as well as the number one as an abstract object.

'fourness' represented. Instead, the four objects occupy separate object files. In this manner, Carey proposes that mental models associated with *n* individuals are associated with the first members of the counting list (up to $n = 4$):

> The meaning of the word 'one' could be subserved by a mental model of a set of a single individual {i}, along with a procedure that determines that the word 'one' can be applied to any set that can be put in 1–1 correspondence with this model. Similarly, two is mapped onto a longterm memory model of a set of two individuals {j, k}, along with a procedure that determines that the word 'two' can be applied to any set that can be put in 1–1 correspondence with this model. And so on for 'three' and 'four'. (Carey, 2009, p. 477)

However, there is one potential problem with this account. Establishing a one-to-one (1–1) correspondence is a relatively sophisticated ability that appears to develop only later in ontogeny. Children who are able to do transitive counting do not necessarily grasp one-to-one correspondence. Empirical studies show that there is a stage of development at which children fail to make the inference from 'these sets have the same number of elements' and 'this set has *n* elements' to 'the other set has *n* elements' (Izard et al., 2014; Sarnecka & Gelman, 2004; Sarnecka & Wright, 2013). Buijsman (2019) has argued that such data show that, even though children are able to recognise that two sets can be put in a one-to-one correspondence, they are not able to make even simple inferences about the cardinality of the sets before they learn the cardinal interpretation of numeral words. Therefore, it seems that, before children possess (at least some) cardinal number concepts, they do not yet properly grasp one-to-one correspondence. In addition, researchers report that the anumeric Pirahã do not successfully complete one-to-one matching for collections of items larger than three (Everett & Madora, 2012; Frank et al., 2008; Gordon, 2004). This is evidence that grasping one-to-one correspondence and even the simpler ability to do one-to-one matching for small collections are developments that require having a system of numeral words in place. Establishing a one-to-one correspondence does not appear to be the kind of universal cognitive ability that proto-arithmetical abilities are.[8]

However, as argued by Beck (2017), establishing the one-to-one correspondence between mental models and sets is not done explicitly at the second stage of the bootstrapping process. Instead, the second stage is due to 'computational constraints' concerning the way the mind processes

[8] It should be mentioned here that Heck (2000) has argued that a cognitively more primitive notion of 'just as many' precedes the ability to do one-to-one matching in ontogeny.

representations in mental models, that is, 'procedures that govern how those representations can be manipulated' (Beck, 2017, p. 116). Therefore the one-to-one correspondence involved in this process is based on constraints of how the mind manipulates the representations in object files, and it is not some cognitive process carried out consciously by the child. As described by Beck: 'While the object files explicitly represent objects, they only implicitly represent that the objects are in one-to-one correspondence' (Beck, 2017, p. 119).

The next question is then how these implicit representations develop into number concepts. Beck argues that children acquire number concepts through 'counting games' in which they learn that the final numeral word in a (transitive) counting sequence is associated with the cardinality of the collection of items. At its simplest, a counting game consists of pointing to each member of a collection while rehearsing the ordered counting list (Beck, 2017, p. 119). Through repetition in games of counting physical items, children thus learn to associate numeral words with cardinalities in a systematic way. In this process, the counting list transforms from a meaningless list of words into a system of numerical representations, by consistently associating the last word on the list with the cardinality of the collection, 'endowing the words in the count list with new conceptual roles' (Beck, 2017, p. 119). As we have seen, this happens gradually as children become one-knowers, two-knowers, etc.

Finally, to complete the bootstrapping process, in a process of inductive and analogous reasoning, at the third stage of bootstrapping, children are able to generalise on this principle beyond the range of the OTS, becoming cardinality-principle knowers. In Beck's account, this stage is carried out as follows: the first four numerals associated with the OTS (at the second stage of bootstrapping) refer to quantities separated by one individual, so consequently the same numerals when associated with counting games are also separated by one individual (Beck, 2017, p. 119). Extrapolating from this, children are inductively able to grasp that every word in the counting list designates a quantity one more than the quantity designated by the previous word. This is the stage when children become CP-knowers, and it completes the bootstrapping process of number concept acquisition (Beck, 2017; Carey, 2009; Pantsar, 2021a).[9]

[9] 'Complete' is admittedly a misleading word to use here, given that a lot of changes still happen after acquiring number concepts, both on the personal and the sub-personal level. For example, evidence shows that, in symbolic processing of numbers, there is a consistent move of brain activity from the frontal regions to the parietal regions between childhood (at twelve years of age) and adulthood (Ansari, 2008).

The bootstrapping account has two great strengths. First, it can explain the empirical data on number concept acquisition in children. Perhaps most importantly, it is consistent with the different stages of becoming a subset-knower, as well as that of ultimately becoming a CP-knower. In addition, it explains why people in cultures without numeral words do not acquire number concepts: because of the lack of counting lists, they do not go through the first stage of the bootstrapping process. The second strength of the bootstrapping account is that it is conceptually coherent and does not make unjustified assumptions. Unlike nativist accounts, the bootstrapping account does not assume innate number concepts. The main factors that are assumed are proto-arithmetical abilities, as well as counting lists and counting games, the existence of which is uncontroversial.

However, even if the bootstrapping account were completely uncontroversial – and I am definitely not claiming that it is – it should not be accepted without argument that the versions of the bootstrapping account presented so far in the literature correctly identify the role of proto-arithmetical abilities and the other factors in the process of number concept acquisition. In the Carey-Beck version of bootstrapping, the OTS is seen as the central (and perhaps only) cognitive core system involved in number concept acquisition. However, I have argued that we should also be prepared to see a role for the ANS in the bootstrapping process (Pantsar, 2021a). This argument is based mainly on the observation that the OTS alone gives no cognitive grounds to process collections greater than four objects in terms of numerosities. Why should the number concepts FIVE, SIX, SEVEN, and so on be so easy to grasp for children growing up in arithmetical cultures if the OTS is the only key cognitive system involved and its limit is the concept FOUR? I have argued that the ANS can play an important role in this, as children at the third stage of the bootstrapping process already have extensive experience of the world in terms of larger numerosities due to approximate numerosity estimations. Unlike the OTS, the ANS is not limited to small numerosities, so we should be open to the idea that the ANS can play a part in the acquisition of larger number concepts. This research is still at early stages and describing the exact roles of the OTS and the ANS in the bootstrapping process demands more empirical data. However, there does not seem to be any reason why this kind of hybrid OTS–ANS model concerning bootstrapping should not be pursued.[10]

[10] Indeed, there is evidence that, with larger numerosities, the approximate numerosity representations are linked to numeral words as children learn to count up to them (Lipton & Spelke, 2003).

But equally importantly, it should be noted that while the bootstrapping account as developed by Carey and Beck emphasises the importance of the OTS, they do not limit the approach to proto-arithmetical abilities. Indeed, already the first step of the bootstrapping process, acquiring the counting list, clearly demands cognitive abilities beyond the OTS. Merely remembering and reciting the counting list demands linguistic abilities that are not present in many subjects that possess proto-arithmetical abilities (e.g., human infants and non-human animals). In addition, being able to engage in counting games demands abilities of social cognition that humans reach only later in childhood, and most non-human animals never acquire. From shared attention to understanding the numeral words and their connection to the bodily actions (such as pointing), engaging in counting games requires a wide range of social abilities.

These observations point us to an important element in the bootstrapping process that I have developed in Pantsar (2021a), namely, that the bootstrapping of number concepts is a *culturally* shaped process. Counting lists are cultural constructs and, as we have seen in cultures without them (like the Pirahã and the Munduruku), people do not acquire number concepts. But the cultural influences do not stop there. The counting games that Beck writes about are also clearly cultural constructs. Indeed, the whole practice of teaching those games along with numeral words is culturally developed.[11] But how do cultural constructs shape the development of our cognitive abilities? The literature on number concept acquisition has mostly been concerned with core cognitive capacities that are ancient and the product of biological evolution. However, as we have seen, we also need to include culturally developed constructs (like counting games and numeral word lists) in accounts of number concept acquisition. Therefore, we will need to move our focus to cognitive capacities that are phylogenetically more recent and the product of *cultural* evolution. This will be the topic of Chapter 3 and Part II of this book.

2.5 Critics of Bootstrapping

While the bootstrapping account of number concept acquisition has received support among researchers, there have also been critical voices.

[11] Consequently, we can expect that cultural differences in counting games are connected to differences in the bootstrapping process. To the best of my knowledge, this hypothesis has not been tested empirically, but it would be an important way to evaluate the Carey–Beck account of bootstrapping.

We have already treated the nativist and ANS-based theories of number concept acquisition, which clearly go against the bootstrapping account. In nativist views, like that of Gelman and Gallistel (1978), number concepts are innate so the developmental stages described in the bootstrapping account must concern something else than number concept acquisition. In ANS-based views like that of early Dehaene (Dehaene, 1997; Dehaene & Changeux, 1993), number concept acquisition is thought to happen (at least to a significant part) independently of object files. If these theories are correct, the bootstrapping process, if it takes place at all, cannot be about number concept acquisition in the sense argued by Carey (2009).

The views of Gelman and Gallistel and Dehaene are clearly incompatible with Carey's bootstrapping account of number concept acquisition. In addition to them, criticism specific to bootstrapping has also been presented in the literature, focusing mainly on two main lines of argumentation, what Beck (2017) calls the *deviant-interpretation* and *circularity* challenges. The deviant-interpretation challenge has been around in philosophy for a long time and it is famous from two particular formulations. The first one is due to Kripke's interpretation of Wittgenstein's rule-following challenge (Kripke, 1982). This is often referred to as the 'Kripkenstein' challenge and it asks how we could ever unassailably know what rule a function (or a sequence, series, etc.) follows if we only know a finite subset of its values. For example, the sequence 1, 2, 3, 4, 5, 6, 7, 8, . . . may suggest that the next number is 9, but how can we know that it is not, say, 1? Instead of the rule being 'add 1', it could be something like 'after 8, start again from 1'. The second formulation of the deviant-interpretation challenge is associated with the 'new riddle of induction' presented by Goodman (1955). In this riddle, Goodman imagines a word 'grue' that has been used as a synonym for 'green' until some point in time, after which it has been used as a synonym for 'blue'. What, then, is the meaning of 'grue'? Although 'grue' is an artificial construction, how can we know that our concepts and words do not change meaning like 'grue' does?

Rips and colleagues have presented a version of the deviant-interpretation challenge specific to number concept acquisition and bootstrapping (Rips et al., 2006). They ask what in the bootstrapping process prevents acquiring, say, a cyclical number system in which, after acquiring the concept TWELVE, the next number concept is ONE instead of THIRTEEN? Rips and colleagues argue that, instead of a bootstrapping process in the way Carey describes it, there needs to be in place a mathematical schema with content similar to the Dedekind–Peano axioms to ensure that a deviant interpretation does not arise (Rips et al., 2008).

This challenge has received different answers. Margolis and Laurence (2008) have argued that, in a nativist view that presupposes stronger innate representational systems for quantities beyond the OTS range, the problem does not arise. While this may be the case, given the problematic nature of nativist notions of number concepts, I find this response less than satisfactory. As we have seen in Section 2.2, there is no evidence of such a strong innate representational system when it comes to exact numbers. Thus, pursuing a rule-following account seems to be the right approach, even given the problem posed by Rips and colleagues.

However, the rule-following problem may not be as serious as it first seems. Beck (2017) has argued that what Rips and colleagues present is in fact a criticism against all inductive learning and not only that of concept learning, in particular learning number concepts. I agree with this point. The rule-following challenge is not in any way exclusive to, or even characteristic of, Carey's account of bootstrapping of number concepts. Whatever rule we posit for learning number concepts, there always exists a chance that the rule deviates at some point. The rule could apply until, say, the number concept related to 5,432,424, but the rule to grasp the numeral could fail with the number concept associated with 5,432,425.

Indeed, there are only two possible scenarios in which this rule-following challenge does not apply. First of these is the position according to which all number concepts are innate. The second is that there is an innate recursive rule that gives us the successor principle. The first option is clearly untenable, given that the brain is finite and the amount of natural numbers is infinite. The second option is possible, but unsupported by evidence. There are cultures that do not have such recursive rules, like the Pirahã and the Munduruku (Gordon, 2004; Pica et al., 2004). In addition, grasping such recursive rules about integers appears to be a relatively late ontogenetic development, only taking place when children have already acquired many number concepts (Pantsar, 2018a; Sarnecka & Carey, 2008).

Of course it is possible that this kind of recursive rule is innate but does not emerge universally, but it is highly problematic to assume that such a recursive rule is the product of biological evolution. There is certainly no evidence supporting that. We cannot know the origin of numeral word systems, but the emergence of recursive written numeral symbol systems is a relatively recent phylogenetic development, dating back to around 5,000 years ago (Fabry, 2020; Pantsar, 2019b; Schmandt-Besserat, 1996). This does not preclude the possibility of older, perhaps innate recursive systems, but there is no evidence of non-literate cultures having recursive numeral systems (Everett, 2017; Ifrah, 1998).

While an innate recursive rule that would give us a mathematical schema with content like the Dedekind–Peano axioms, as argued for by Rips et al. (2008), is unsupported by evidence, it is possible that there is some kind of innate *bias* toward such a recursive rule. However, the empirical evidence only suggests the existence of two proto-arithmetical abilities that could be the source of that bias, and neither subitising nor ANS-based estimating prevents the possibility of deviant rules. This should prompt us to look for the roots for recursion elsewhere. It has been suggested that sensitivity to rhythm and thus regularity could be a factor in grasping recursivity (Quinon, 2021). I believe that this possibility should be pursued, but it seems that the most likely source of recursivity in ontogeny comes from *cultural* and not any biologically evolved, innate factors.[12]

Indeed, these cultural factors are already present in the bootstrapping process as discussed in Section 2.4. The counting list, which is acquired at the first stage of the process, is a cultural construct that (in many cultures) shows recursivity. This takes different forms in different languages, and the particular form appears to have an effect on the way children acquire number concepts and arithmetical skills. For example, while an English speaker needs to learn the new words 'eleven' and 'twelve', in Mandarin the corresponding words are 'shí-yī' and 'shí-èr' (literally, ten-one and ten-two, respectively). Generally, the Mandarin numeral word system has shorter words and follows the recursive decimal structure more clearly than the English one. Miller et al. (1995) have proposed that these factors can explain why native Mandarin speakers learn to count faster, which would imply that the numeral word system they are introduced to affects the learning trajectory in grasping the numeral symbol system and acquiring number concepts.

In addition to the counting list, also the counting games mentioned by Beck are a cultural product. If the numeral system were cyclical, surely this would be present in some way in the counting games that children are acquainted with. But since the numeral system is linear and recursive, the children do not acquire a cyclical number system. In Chapter 3 we will treat this question in more detail, but it seems highly likely that the immediate source of acquiring a linear number system in ontogeny is not some mathematical schema (whether innate or acquired). It is simply that children are *taught* a linear number system, with no loops of the type

[12] In Part II of this book, we will see how this question can be approached in terms of phylogeny and cultural history.

suggested by Rips et al. (2006). However, this does not mean that proto-arithmetical abilities are not involved in this process. They can play an important part in explaining how linear – and not cyclical – number systems have emerged in phylogeny and cultural history. Indeed, I will provide such an argument in Part II of this book.

Proto-arithmetical abilities can also play another role in ontogeny. For example, as argued in Pantsar (2021a), the lack of loops in number systems can be (partly) explained by the influence of the ANS. While the OTS only concerns small numerosities, there is no such limit for the ANS. By applying the ANS to larger and larger numerosities, a child can develop a sense that there is no limit to how big numerosities can be. This could be the source of an early grasp of the idea that numbers form a continuing progression without loops and, together with the OTS, it could explain why such a progression is linear instead of logarithmic.[13]

In addition to the deviant-interpretation challenge, the bootstrapping account has also faced the circularity challenge. This challenge was most famously formulated by Fodor, who wrote that:

> There literally isn't such a thing as the notion of learning a conceptual system richer than the one that one already has; we simply have no idea of what it would be like to get from a conceptually impoverished to a conceptually richer system by anything like a process of learning. (Fodor, 1980, p. 149)

Fodor's rejection of the notion notwithstanding, clearly this kind of acquisition of a conceptually richer system is happening in number concept acquisition according to the bootstrapping account. Through a process of learning, we are acquiring a conceptually richer system than the one we previously had. Unsurprisingly, this resulted in criticism from the Fodorian camp. Specifically against the bootstrapping account, Fodor and his followers have argued that any such new concepts would need to be possessed by the learner already (Fodor, 2010; Rey, 2014). More specifically, as argued by Rey (2014, p. 117), Carey's notion that successive numerals are connected by the relation 'one greater than' (Carey, 2009,

[13] This is an answer to what I have called the *stronger* deviant-interpretation challenge (Pantsar, 2021a). Since the ANS-based estimations are logarithmic, why do we not bootstrap a logarithmic number system? The Carey–Beck answer is that the ANS does not play a role in the bootstrapping process, or at least not an integral one. But then how do we get the initial idea that there are also numerosities beyond the OTS range? I have argued that ANS can be the source of this idea, while the OTS can be the source of the idea that numbers form a linear progression (for the detailed argument, see Pantsar, 2021a).

p. 277) appears to assume that there is a pre-existing grasp of 'one greater than', which is then used to define the successor relation.

While he sees the deviant-interpretation challenge as a general problem in inductive learning, Beck (2017) sees the circularity challenge as more worthwhile. His response (p. 119) is based on the internal 'computational constraints' that the mental object files impose on the bootstrapping process. The computational constraints associated with the OTS allow grasping that the numeral 'one' is associated with one object file being occupied, thus acquiring the number concept ONE. During the learning process, a child then learns that combining a collection of 'one' *F* with 'one' *F* creates a collection of 'two' *F*s, acquiring the concept TWO. This happens until four *F*s give the concept FOUR, which is the limit of the OTS. In the next step of the bootstrapping process, the child then uses analogy and induction in grasping the extrapolation that words in the counting list generally work in a similar manner, namely, that the next numeral on the list is associated with the process of combining the previous collection of *n* *F*s with 'one' *F*. Beck argues that this account avoids the circularity challenge. The argument is that, by the time the analogous and inductive process of the third stage of bootstrapping takes place, the placeholder words in the counting list have already been partially interpreted due to the computational constraints. The child has already grasped that quantities designated by subsequent numeral words from 'one' to 'four' differ by 'one' individual, and this realisation can be used to infer that the same difference of 'one' also characterises the rest of the numeral list (Beck, 2017).

Does Beck's argument manage to disarm the circularity challenge? I am quite confident that if it does not already, the argument can be further developed to meet all forms of the circularity challenge. The exact mechanism through which the bootstrapping process takes place will need to be detailed and tested by empirical studies, but the general idea is highly plausible and makes a strong case against circularity. But equally importantly, I do not see the circularity challenge as the kind of behemoth that it sometimes seems to be in the literature on learning. The reason for this is simple: in my view, the only account that does not face the circularity challenge, namely the nativist account of concepts of Fodor, is altogether implausible. Nowhere is this more evident than in the context of mathematics. Mathematics is full of abstract concepts that are the product of a long line of development, invention and innovation. In addition to there being no evidence to suggest that these concepts exist innately in the human mind, even the theoretical possibility of that being the case seems far-fetched. At the very least, the burden of proof should be on the other side. If acquiring novel concepts truly is impossible, this should be supported by evidence and direct argumentation.

While I do not think that there are at present reasons to generally doubt the bootstrapping account, this does not mean that it is in its current form conclusive and the question of number concept acquisition should not be open to other approaches. Schneider et al. (2022), for example, have reported that, even after acquiring the cardinality principle, children (of five years of age) may still not consistently match sets of small cardinalities. They argue that, in addition to understanding the meaning of numeral words, in order to master the notion of exact equality, there also needs to be a deeper knowledge of a counting system. Thus, it is possible that bootstrapping can only explain the acquisition of number concepts, but not *applying* number concepts to processes such as establishing exact equality.

I think that it is important to recognise this limitation of the bootstrapping account. Acquiring number concepts is only the first stage in learning arithmetic and it would be mistaken to believe that children automatically become competent in applying number concepts more generally. The importance of understanding counting processes for acquiring arithmetical skills and knowledge has been known for a long time (see, e.g., Fuson, 1987). Therefore, while I endorse the bootstrapping account, we must recognise its role in explaining the development of arithmetical skills and knowledge in ontogeny. As we will see in Chapter 3, it is also important to place the bootstrapping account in a wider theoretical framework of how humans learn new contents and practices and acquire new concepts. It is only after this work is done that we can properly assess the bootstrapping account. However, in this regard it is no different from any other account of number concept acquisition. Indeed, as I aim to show in Chapter 3, the bootstrapping account fits well with a particular framework of learning that emphasises our place as learners situated within cultures.

To conclude this chapter, the topic of animal ability in what appears to demand number concepts should be discussed. Perhaps most famously, Pepperberg (2012) reported statistically significant accuracy by a grey parrot in addition tasks, conducted with objects, English numeral words and Indo-Arabic numerals. The addends were not visible when the parrot was given the task, so, unlike many other tasks, which can be explained by the application of proto-arithmetical abilities, the parrot seemed to be doing something more. Namely, it seemed to possess some kind of manipulatable numerosity representations. Boysen and Berntson (1989) reported similar activity in chimpanzees. Even though both parrots and chimpanzees can learn the meaning of symbols and words, it would be questionable to say that they have linguistic ability in the sense humans have. But, in the bootstrapping process, acquiring numeral words plays a crucial role. Is this evidence against the bootstrapping account?

This would be a mistaken conclusion based on the reported studies. In both the parrot and the chimpanzee studies, the animals were trained with numeral words and numeral symbols. Thus, what they were doing already goes beyond their proto-arithmetical abilities. That parrots and chimpanzees can perform so well (although not without errors) in addition tasks is remarkable. But it does not show that they possess innate arithmetical abilities. What it does show, I believe, is that these animals show intelligence and quite a sophisticated ability to learn and associate words with objects and properties. In this process, it is at least plausible that they acquire genuine number concepts (for an argument that they do, see Nelson, 2020). But it is important to remember that the abilities they developed were taught to them by humans who were arithmetically trained. We should be open to the possibility that some non-human animals can acquire number concepts and learn some basic arithmetic. Indeed, perhaps their learning process would be different from the bootstrapping process. However, to the best of our knowledge, this does not happen without a training process guided by a human agent. In this way, the results on animal ability with addition are consistent with the bootstrapping account of number concept acquisition.

2.6 Summary

In Chapter 2, I focused on the acquisition of number concepts related to natural numbers. I reviewed nativist views, as well as Dehaene's early view that number concepts arise from estimations due to the approximate numbers system. I ended up focusing in most detail on the bootstrapping account of Carey and Beck, according to which the object tracking system is the key cognitive resource used in number concept acquisition. However, I endorsed a hybrid account that also includes an important role for the approximate numerosity system. I then reviewed some of the criticism against the bootstrapping account, concluding that, while more empirical data is needed to establish its correctness and details, currently it provides the most plausible account of early number concept acquisition.

CHAPTER 3

Enculturation

In Chapter 2, we have seen how children can acquire number concepts in their ontogeny. I ended up endorsing a bootstrapping account along the lines presented by Carey (2009) and Beck (2017) as the direction to pursue. But, unlike nativist views, the bootstrapping account clearly includes cultural elements, such as numeral word lists and counting games. How can an ontogenetic account of number concept acquisition accommodate such culturally shaped learning? In this chapter, I will tackle that topic through the notion of *enculturation*. First, however, it is important to make some conceptual clarifications with regard to the way the human brain is suited for learning mathematics. Clearly our brains allow for the development and learning of mathematics. But does it make sense to claim, as often is done, that our brain *is* mathematical?

3.1 Mathematical Brains vs. Brains for Mathematics

In the Introduction, I stressed how important it is for the present approach to use terminology that is coherent and not misleading in terms of the cognitive capacities that are assumed. These included, for example, the conflation between numbers and numerosities, as well as between proto-arithmetical and arithmetical abilities. Similarly, the literature is rife with questionable statements concerning what the cognitive domain of arithmetic, and mathematics in general, is thought to be. Instructive of this are the two different titles that the widely read book by the neuropsychologist Brian Butterworth (1999) on numerical cognition received on the two sides of the Atlantic. In New York, the book was published with the title *What Counts*, with the subtitle *How Every Brain Is Hardwired for Math*. In the wordplay of the main title, one is prompted to the question of what exactly is the agent with which counting and other arithmetical abilities should be associated? This is a *bona fide* question. As we will see, many researchers disagree with the brain-centric view and emphasise the embodied aspect of the development of arithmetical cognition (Bender & Beller, 2012; Fabry, 2020; Menary, 2015; Overmann, 2016). A typical example of this is the role of finger counting in learning numeral words. But this already suggests that Butterworth's subtitle is inaccurate. His idea is that mathematics is based fundamentally on the structure of our brain, which is the result of evolutionary adaptations. But if finger-counting is integral to learning arithmetic, it is not just the brain, but also parts of the rest of the human body that is 'hardwired for math'.

However, perhaps a more serious problem is found in the title of Butterworth's book as published in London: *The Mathematical Brain.* In that title, it is not merely claimed that the brain is *hardwired* for mathematics, but that the brain itself *is* mathematical. This is an important difference, since the claim that the brain is hardwired for mathematics is much weaker. A computer is hardwired to run, say, a word processing software but, without such software installed, the computer is not a word processing tool. Analogously, the brain is clearly structured in a way that makes mathematics possible but, without the 'software' for mathematics in place, the brain is not mathematical. However, as we have seen already, the 'software' for mathematics does not come pre-installed, so to speak, in the brain. Natural number concepts already give us an example of mathematical cognitive content that needs to be acquired in ontogeny through culturally developed processes, applying culturally developed cognitive tools and practices (such as numeral words). While the brain appears to be hardwired for mathematics in the sense that there is a universal human capacity for learning mathematics, it seems highly misleading to say that the brain itself is mathematical.

The first observation we should make related to that topic is that there is a difference between something being mathematical and it being describable in mathematical formalisms. The Rubik's Cube, for example, is clearly a physical object. There is nothing essentially mathematical in the plastic that makes up the toy, or at least not any more than there is in any physical object. Yet the Rubik's Cube as a puzzle can be fully modelled mathematically – as is indeed often done in introductory classes of an area of mathematics called group theory. It is an open question how the brain can be modelled mathematically – or indeed if it can be. But if it can, this would not amount to showing that the brain is mathematical, any more than showing that the Rubik's Cube can be modelled mathematically establishes it as a mathematical object.

Importantly, however, the question whether the brain can be (fully) modelled mathematically is generally not seen as a key issue in the study of cognition. The brain as an organ is incredibly complex and a full mathematical model of a human brain might be impossible already due to the complexity involved. Instead, in the study of cognition the more pertinent question is whether the *mind* can be modelled mathematically. Granted, in addition to empirical researchers, there are also philosophers who do not think that this distinction is crucial. According to the mind/brain identity theory, cognitive processes are only brain processes (Smart, 2017). It is not possible to enter that debate here, but there are several ways in which identity theory has been challenged (see, e.g., Chalmers, 1997). The one that will interest us most in this book is the argument that the mind does not reduce to the brain since it is *embodied* (Varela et al., 1991). According to the embodied cognition thesis, the cognitive agent's physical body and its interactions with the environment are significant for cognitive processes (Shapiro & Spaulding, 2021). In this way, the embodied mind cannot be reduced to the brain.

As we will see, the embodied cognition thesis is an interesting topic in the development of arithmetical cognition. For now, however, the important message to take home from research on embodied cognition is that we cannot equate the brain and the mind. But, even if we accepted the mind/brain identity theory, it is misguided to say that the brain (or the mind) does mathematics, in the sense of carrying out mathematical procedures. What carries out the mathematical procedures is the *person*, that is, the cognitive agent.[1]

With these conceptual clarifications in place, we can see the problem with the view that the human brain is fundamentally mathematical. I have used the titles of Butterworth's book as an example but discussing the 'mathematical brain' is a much wider and enduring problem in the literature. To give just one example, a recent special issue of the journal *Developmental Cognitive Science* was titled 'Advances in Understanding the Development of the Mathematical Brain' (Hyde & Ansari, 2018). Far from being merely a terminological problem in the early literature, the idea of the mathematical brain has endured, as misguided as it is. As we have seen, even basic arithmetical abilities are not present in the human brain unless they are learnt in a suitable cultural context. From what the current

[1] This distinction is often not made in the cognitive sciences and it is called the 'mereological fallacy' (Bennett & Hacker, 2003, p. 72; Metzinger, 2013, p. 7). For example, Dehaene's latest book is titled '*How We Learn: Why Brains Learn Better Than Any Machine . . . for Now*' (Dehaene, 2020). With the present terminological distinctions, I would paraphrase the subtitle as 'Why Humans Learn Better Than Any Machine'. However, as far as the scientific content is considered, there is no reason to believe that Dehaene's use of terminology causes it to be incompatible with the present approach.

state of art in empirical research tells us, the brain is not mathematical in any feasible sense. Mathematics is a *cultural* phenomenon and cultural phenomena cannot be conflated with properties of the brain.[2]

Of course Butterworth (1999) was already well aware of the limitations of what are in this book called proto-arithmetical abilities. Indeed, his account is open to cultural influences. What he argues for is an innate 'number module', which functions as the cognitive basis for arithmetical abilities. But to go beyond the subitising range '... we need to build onto the Number Module, using the conceptual tools provided by our culture. We add these to the other items of our conceptual toolkit through learning from other people' (Butterworth, 1999, p. 7). The number module should of course be called the *numerosity* module and, as we will see, there are problems involved in postulating such mental modules. But, in a more general sense, I agree with Butterworth's approach. There are evolutionarily developed proto-arithmetical abilities (related to what he calls the number module) that are employed in the development of arithmetic. In this development, cultural factors play an important role.

The other author of an influential book on mathematical cognition, Stanislas Dehaene, would seem to agree, even though he also comes to the conclusion based on questionable conceptual work. In the *precis* for his book *The Number Sense*, Dehaene writes: 'In the course of biological evolution, selection has shaped our brain representations to ensure that they are adapted to the external world. I have argued that arithmetic is such an adaptation' (Dehaene, 2001a, p. 31). But clearly arithmetic is *not* such an adaptation; *proto-arithmetic* is. Similarly, Butterworth's subtitle that the brain is hardwired for mathematics is acceptable, but the title that the brain *is* mathematical is not. However, like Butterworth, Dehaene expands on his account in a much more acceptable way:

> Specific to the human species, however, is a second level of evolution at the cultural level. As humans, we are born with multiple intuitions concerning numbers, sets, continuous quantities, iteration, logic, or the geometry of space. Through language and the development of new symbols systems, we

[2] In this sense, I am much fonder of the title of Andreas Nieder's recent book, *A Brain for Numbers*. However, its subtitle, *The Biology of the Number Instinct*, goes astray. I do not believe that there is a number instinct, even though our brain with its proto-arithmetical capacities is fit for developing cognition of numbers. These proto-arithmetical capacities may be called the *numerosity instinct*. This difference is not merely terminological and reflects the different emphasis Nieder and I put on biologically and culturally evolved factors in the development of numerical cognition, even though we both see an important role for both.

> have the ability to build extensions of these foundational systems and to draw various links between them. (Dehaene, 2001a, p. 31)

Of course, I cannot agree with this quotation in everything. I do not believe that we are born with intuitions concerning numbers or any other mathematical objects or processes. We are born with proto-arithmetical abilities that can in a proper cultural setting *lead* to possessing something that can be called 'intuitions' about numbers and other mathematical objects. It is also possible that we possess similar proto-geometrical, as well as possibly proto-set-theoretical and proto-logical abilities. However, it is important that all these are distinguished from any genuinely mathematical intuitions.

When it comes to the mathematical intuitions, however, Dehaene points to the crucial aspect. Language and symbol systems allow extending beyond proto-mathematical abilities. We will see that there are also other factors involved, but Dehaene seems to be fundamentally correct about the most important aspect: it is evolution on the *cultural* level that we need to understand in order to explain how proto-mathematical abilities can develop into mathematical knowledge and skills. In Part II of this book, I will study the matter of cultural evolution in detail in the context of arithmetic. However, before that, we need to understand how culturally developed factors can influence our cognitive development. Clearly this happens all the time, but *how* does it happen? If Butterworth and Dehaene – as well as Carey, Beck and the present account – are on the right track, there needs to be a feasible way in which cultural factors enable extending and transforming our proto-arithmetical abilities into arithmetical abilities. This is the question we will tackle next.

3.2 What Is Enculturation?

The notion of enculturation I want to pursue in this book is based on the one originally developed by Richard Menary (2014, 2015). The main motivation to introduce the framework of enculturation is to include both culturally and evolutionarily developed aspects in the development of cognition. In short, enculturation refers to the transformative process in which interactions with the surrounding culture shape the acquisition and development of cognitive practices (Fabry, 2018; Menary, 2015; Pantsar, 2019b). Through the neural plasticity of the brain, which enables both structural and functional variations, new cognitive capacities can be acquired in culturally specific contexts (Ansari, 2012; Dehaene, 2009). Menary calls this mechanism *learning driven plasticity* (LDP) (Menary, 2014).

The central idea of Menary's conception of enculturation and learning driven plasticity is based on Dehaene's notion of *neuronal recycling*, according to which it is possible to acquire culturally developed cognitive abilities like reading and writing by redeploying older, evolutionarily developed neural circuits for new culturally specific functions (Dehaene, 2009; Menary, 2014). Michael Anderson has argued that, instead of neuronal recycling, the basic organisational principle of the brain is that of more general *neural reuse* (Anderson, 2010, 2015). We will return to this topic in Section 3.4, but for now we only need to accept the uncontroversial idea that learning in a cultural context can result in some kind of reuse of neural resources. Aside from reading and writing, the enculturation framework has been pursued to understand the development of mathematical cognition (Fabry, 2020; Jones, 2020; Menary, 2015; Pantsar, 2019b, 2020). Much of the work on enculturated mathematical cognition has focused on the acquisition of arithmetical abilities (Fabry, 2020; Jones, 2020), but I believe that enculturation can help us understand generally the emergence of mathematical cognitive practices. In Pantsar (2021d), for example, I propose how to apply the enculturation framework in explaining the development of geometrical knowledge. Similarly to the case of arithmetical knowledge, also geometrical cognition appears to be (at least partly) based on proto-mathematical abilities (Hohol, 2019; Spelke, 2011).

Indeed, one of the strengths of the enculturation account in the present context is how it can unify the development of different cognitive capacities under one theoretical framework. The framework itself does not demand any specific assumptions concerning mathematical cognition. Rather, it can help provide a general account of culturally driven learning, which can then be applied to particular fields of cognition. In the case of arithmetical knowledge, the strength of the enculturation framework is obvious. Whether we favour neuronal recycling or neural reuse as the organisational principle, it is highly likely that the development of arithmetical cognition involves some kind of application of proto-arithmetical abilities. The enculturation framework can then help us understand how proto-arithmetical abilities, through learning in a specific cultural context, can be evoked in acquiring arithmetical abilities.

3.3 Enculturated Acquisition of Number Concepts and Arithmetic

In the case of arithmetical knowledge, the first question to ask in the enculturation framework is how number concepts are acquired. This question is a prime example of a phenomenon that the notion of enculturation can potentially help us understand. Proto-arithmetical abilities, as we have established, do not involve number concepts. In the first years of our lives, we have abilities with numerosities, but we do not possess any number concepts. Somehow, quite possibly through a bootstrapping process, in most cultures children in the span of a few years turn from having no number concepts into possessing a large amount of number concepts, with the tools in place to gain more. Indeed, children at some point acquire a *general* concept of a natural number. This does not happen without the appropriate cultural surroundings for acquiring number concepts, as shown by the cases of the Pirahã and the Munduruku (Gordon, 2004; Pica et al., 2004). But with a conducive cultural setting in place, most neurotypical children acquire number concepts with relative ease. This trivial-sounding observation is of utmost importance in explaining number concept acquisition. The cultural setting is absolutely integral to acquiring number concepts, and subsequently to acquiring arithmetical skills and knowledge. In explaining the development of arithmetical cognition, this makes cultural factors a central research topic.

As important as cultural factors are, it would be equally misguided to dismiss the importance of proto-arithmetical abilities for the acquisition of number concepts and arithmetical knowledge and skills. This is the strength of the enculturation framework. It does not make the classic nature vs. nurture distinction between genetically determined and culturally shaped influences when it comes to the development of cognitive abilities. Rather, learning is seen inherently as a process in which interactions with the surrounding environment shape the way cognitive abilities develop. Accounts of this process are often focused on the brain, as in Menary (2015), but it should be seen as a wider phenomenon. To complement Menary's notion of learning-driven plasticity, Regina Fabry has introduced the notion of *learning-driven bodily adaptability* (LDBA) to describe the way that culturally shaped learning processes are not restricted to the brain. Similarly to the way the learning-driven plasticity of the brain leads to the emergence of new neural circuits, learning-driven bodily adaptability leads to the emergence of new bodily action patterns (Fabry, 2018, 2020; Fabry & Pantsar, 2021).

In accounts of number concept acquisition, both LDP and LDBA should be included. One key reason for this is the importance of fingers for early counting processes. Most cultures appear to have widely shared practices for finger counting and children across cultures use fingers for early counting and calculation procedures (Barrocas et al., 2020; Bender &

Beller, 2012; Butterworth, 1999; Fuson, 1987; Ifrah, 1998). Finger gnosis (or finger gnosia), the ability to differentiate between one's own fingers without visual feedback (Wasner et al., 2016), has been shown to be a predictor of numerical and arithmetical ability levels both in children and adults (Noël, 2005; Penner-Wilger et al., 2014; Wasner et al., 2016). Fabry (2020) has argued for the importance of LDBA in explaining this connection between finger gnosis, finger counting and the development of arithmetical abilities. She has also stressed how the realisation of finger counting can impact the neuronal realisation of arithmetical abilities, thus pointing to an 'important reciprocal influence of LDBA and LDP in the course of enculturation' (Fabry, 2020, p. 3704).

How does this account of enculturation fit with the bootstrapping account presented in Section 2.4? Recall the importance that Beck (2017) saw for external computational constraints in the process, including counting games. I see the enculturation framework as a perfect setting for explaining how such counting games influence learning. Counting games are, obviously, culturally developed practices. When children learn them, they act as embodied agents in their environment. This concerns in an important way (but is not limited to) their use of fingers. Aside from finger counting, early counting processes are also associated with pointing (Wilder et al., 2009). Counting games, aside from eye-movements – which can be seen as a bodily process in its own right – often also involve pointing processes. For example, in counting games a child typically points to one object, utters a numeral word, and then points to another object, uttering the next numeral word on the list (Fuson, 1987). This is culturally learned behaviour that makes it possible to associate the counting list with collections of discrete objects. Typically, through shared attention with a teacher (or parent or someone else), the child learns the behaviour because they see someone else, someone they inherently trust, act out that practice. When a parent points in turn to three different objects and utters the numeral word list 'one, two, three', the child will focus on this practice. This ability and propensity for shared attention develops very early in infancy, giving vital help in the learning process (Adamson & Bakeman, 1991).

Without the required culturally developed practices in place, children do not acquire number concepts. A good example of this is the fact that the Pirahã do not consistently complete one-to-one matching tasks for collections larger than three (Everett & Madora, 2012; Frank et al., 2008; Gordon, 2004). While it might be natural to think that establishing one-to-one matching between collections does not require number concepts (which the Pirahã do not possess), it seems that the two abilities are tightly connected. It has been reported that bi-lingual members of the typically anumeric Munduruku tribe learn to count and acquire number concepts without problems (Gelman & Butterworth, 2005; Gelman & Gallistel, 2004). While this, correctly, has been used as an argument for the close connection between available linguistic resources and the development of numerical abilities, it is important to note that, aside from being exposed to another language (Portuguese), the bi-lingual Munduruku were also exposed to cultural practices of the Portuguese-speaking Brazilians, including counting practices such as counting games.

Aside from culturally developed cognitive practices, an important element in processes of enculturation are cognitive *tools* (Fabry & Pantsar, 2021; Menary & Gillett, 2022). In the case of numerical cognition and arithmetic, these include pen and paper, abacus, electronic calculators and many other widely used tools. But the cognitive tools are in place already in the early counting games. Perhaps the Pirahã difficulty with one-to-one matching, for example, is already partly due to not being enculturated with the kind of objects in early age that children in arithmetical cultures regularly confront (e.g., blocks of wood, marbles and other more or less identical artefacts).

In the above manner, I believe that the framework of enculturation can help complete the Carey–Beck bootstrapping account. Indeed, the present approach seems to be compatible with the writings of Carey and Beck. Carey, for example, writes that 'the capacity to represent the positive integers is a cultural construction that transcends core cognition' (Carey, 2009, p. 287). But how do cultural constructions contribute to cognitive capacities in the ontogeny? Beck (2017) writes about counting games but does not provide a theoretical framework for how cultural factors can shape our cognitive processes. My proposal here is that the framework of enculturation provides the theoretical basis for an account which can be developed in light of new empirical data. Counting games shape the development of our numerical abilities because the plasticity of our brains and the adaptability of our bodies shapes the way our core cognitive proto-arithmetical abilities develop as a result of the interaction with our environment involved in the game. Thus, learning in a cultural context is key to the development of proper numerical and arithmetical abilities.

As we have seen in Section 2.5, the cultural context is crucial both to acquiring a recursive numeral list and being enculturated in counting games in which numbers are learned in a linear order. Indeed, the main criticisms toward the bootstrapping account, as presented in that section, are answered by the enculturation account. Deviant interpretations in acquiring number concepts are avoided because we are enculturated with linear number systems. The circularity of learning, as criticised by Fodor (1980), is avoided because new concepts are developed culturally and can be introduced through processes of enculturation.

Equally important for recognising the cultural influences on number concept acquisition is to note that the enculturation framework can also accommodate the core cognitive basis for the bootstrapping process. Enculturation does not imply that number concepts and arithmetic are purely cultural constructs, in the sense that they are not shaped by our evolutionarily developed cognitive abilities. Indeed, one main feature of the enculturation framework is that it can incorporate evolutionarily developed neural circuits in the learning of culturally shaped abilities. This is exactly what has been proposed in the case of the Carey–Beck bootstrapping account. Evolutionarily developed proto-arithmetical capacities are thought to provide the fundamental cognitive basis for observing the world in terms of numerosities. In the account of Carey (2009) and Beck (2017), the proto-arithmetical ability in question is the OTS. In my own account, both the OTS and the ANS play a role (Pantsar, 2021a). However, in both accounts, proto-arithmetical abilities can give rise to the emergence of number concepts only with access to proper cultural practices. This is consistent with the idea of enculturation.[3]

An important thing to note about the enculturation account is that it does not suggest a single cognitive path for the development of numerical cognition, with uniform characteristics across cultures. Thus, it is possible that, in different cultures, children acquire number concepts in highly different ways. It could be that the counting games used in early numerical education, for example, have different foci in different cultures. This influence can make a difference in terms of which proto-arithmetical abilities are involved, and how.[4] In general, the enculturation account is consistent with the idea that, in number concept acquisition, there may be different emphases on, and perhaps even different orders of, cognitive influences and notions. One such potential difference concerns the notions of one-to-one correspondence and counting, as discussed in Section 2.4. In the bootstrapping account, it was argued, those two notions can both be integral to the development that allows for number concept acquisition.

[3] In Pantsar (2021a), I argue in more detail for the position that the enculturation framework provides the best theoretical platform for the bootstrapping account.

[4] To the best of my knowledge, there have been no empirical studies of this.

What the enculturated bootstrapping account emphasises is that this development and the importance of the two notions, one-to-one correspondence and counting, may vary across cultures.

The way in which the learning of new cognitive practices and the acquisition of new concepts happens on the neuronal level is of some dispute, and will be the topic Section 3.4. For now, however, we should ask how the enculturation account can help explain the development beyond number concept acquisition, to proper arithmetical skills and knowledge. In one way, this question is more straight-forward than number concept acquisition. Indeed, the acquisition of arithmetical knowledge and skills has been the main topic of research on enculturation in mathematics (Fabry, 2020; Jones, 2020; Menary, 2015; Pantsar, 2019b). The short answer is that arithmetical practices are acquired in ontogeny as the result of structured learning, which shapes cognitive practices by establishing new neural circuits and bodily motor patterns. Learning addition through finger calculation, for example, is a natural follow-up to finger counting in this sense. By observing and trying finger calculating processes in a supervised learning setting, children learn to acquire the proper practices.

Later, a similar process is carried out in learning symbol manipulation with pen and paper, for example with addition and multiplication algorithms. The details of how this happens are something that will slowly be revealed through empirical research, but we should expect cultural differences.[5] Tang et al. (2006) reported an fMRI study according to which native Chinese speakers show higher visuo-premotor association of numbers than native English speakers in mental calculation tasks. While native Chinese speakers show increased activity in the premotor association area and the supplemental motor area, native English speakers show higher activation levels in the left perisylvian cortex, which is associated with language processing. Tang and colleagues hypothesise that the wider use of the abacus and the shorter Chinese numeral words are factors in this. These data are consistent with the enculturation framework because they imply that cultural practices (such as the use of cognitive tools and languages) shape the development of arithmetical abilities. It seems plausible that future research will establish many similar differences.[6]

[5] We should also expect differences between different arithmetical operations. Studies show that subtraction, for example, shows greater activity of the intraparietal sulcus (IPS) than multiplication (Prado et al., 2011). This is not surprising, given that the IPS is associated with proto-arithmetical numerosity comparison tasks and subtraction appears to be more closely related to comparison than multiplication is.

[6] For more on this topic, see Fabry and Pantsar (2021).

However, differences between cultural practices can be small and blood flow detection techniques such as fMRI (functional magnetic resonance imaging) may not be able to detect their effects on the brain. Sometimes, as in the experiment of Tang and colleagues, it is possible to detect systematic differences in activation patterns in members of different cultures. But it could also be that some differences in cognitive processes are not connected to differences in activation patterns detectable by techniques like fMRI. This would not mean that processes of enculturation are not present in acquiring those cognitive abilities. The fundamental starting point in developing the enculturation framework is that all learning takes place in an environment that we interact with. Therefore, we should not think of cultural factors as being one influence in the way learning takes place. Cultural factors are *always* present in environments that we share with other humans, and we cannot conceive of learning outside of them (aside from highly unusual cases, like children growing up among animals). Against this background, the key question is not whether enculturation happens but *how* it happens.

3.4 Neuronal Recycling and Neural Reuse

The basic assumption behind the enculturation framework is that there exists a way in which learning transforms our cognitive capacities, which in turn requires that the brain (and the body) has sufficient plasticity to allow for such transformations. Both of these assumptions have been confirmed by many experiments. Monkeys have been trained in both tool use (Obayashi et al., 2001) and symbol learning (Srihasam et al., 2012) and fMRI scans established both increased activity in the associated brain regions and the formation of new activity-sensitive regions in brains. Numeral and letter symbols, in particular, are something that monkeys would never naturally encounter, yet there was a neuronal mechanism in place that the monkey brain applied in symbol learning. This kind of learning changes the anatomy of the brain in a significant way. In the expert tool user macaque, for example, a 23 per cent increase in thickness of the specific region of the cortex associated with the tool use – the anterior parietal region – was reported (Dehaene, 2020, p. 91).

Similar changes in the neuronal structure have been detected in humans in structural MRI scans in many contexts, ranging from expert musicians (in the corpus callosum; Schlaug et al., 1995) to taxi drivers (in the hippocampus; Maguire et al., 2000, 2003). Both the thickness of the cortex and the strength of the cortical connections are improved through constant activity, consistent with one of the basic principles of learning: the more often the neurons in some brain region fire, the more synaptic connections there will be. This transformation of the brain is not limited to synapses and the dendrites and axons in neurons, but also involves the surrounding glial cells (Dehaene, 2020, p. 92).

Standardly, the plasticity of the brain is divided into two phenomena. *Structural* plasticity refers to the way the neuronal connections of the brain can change, while *functional* plasticity refers to the way the function of neurons can change (Chang, 2014; Mateos-Aparicio & Rodríguez-Moreno, 2019). Thus, structural plasticity is associated with changes in neuronal connections within and across brain regions (e.g., changes of grey matter proportions), while functional plasticity is associated with changes in the neuronal correlations of events in different brain regions (e.g., one region of the brain taking over a new function in a damaged nervous system) (Fabry, 2020; Freed et al., 1985; Petersen & Sporns, 2015). How are these forms of plasticity of the brain realised? Currently, there are two competing proposals for the basic organisational principle, both of which have also been argued for as the basis of enculturated acquisition of arithmetical cognition in ontogeny, based on the proto-arithmetical abilities. Dehaene (2009) has argued for the *neuronal recycling* hypothesis, according to which new culturally developed cognitive abilities can be acquired through recycling neuronal circuits originally developed (evolutionarily) for different purposes. Thus, the ability to read, for example, is made possible by recycling circuits developed to perform similar functions, such as visual processing and language comprehension (Dehaene, 2009). Menary (2014) adopted Dehaene's neuronal recycling hypothesis as the organisational principle of his enculturation account, and applied it specifically to the case of mathematical cognition (Menary, 2015). His central idea is that neuronal circuits originally developed for proto-arithmetical cognition are (partly) recycled when learning arithmetic in a cultural setting.

There is evidence in support of this hypothesis. Studies show that the same brain regions (the prefrontal and posterior parietal lobes, especially the intraparietal sulcus) activate when dealing with symbolic and non-symbolic representations of numerosities (Nieder & Dehaene, 2009). It is also established that in adult humans proto-arithmetical abilities are activated also in the context of symbolic numerical tasks, suggesting that the brain regions associated with proto-arithmetical abilities are recycled when learning symbolic arithmetic (Butterworth, 2010; Dehaene, 2011; Spelke, 2011). It has also been established that the corresponding areas in the prefrontal and posterior parietal lobes activate also in the brains of non-human primates, and monkeys and college students show similar activation patterns in association with number ordering and quantity estimation tasks (Cantlon & Brannon, 2007; Piazza et al., 2007). These findings indicate that there is a 'proto-arithmetic circuit' that is recycled for arithmetical purposes in the enculturation process. As we have seen, there are disagreements over which proto-arithmetical system is more important in this process, some arguing for the ANS (Dehaene, 2011; Piazza, 2010) and others for the OTS (Beck, 2017; Carey, 2009), while yet others argue for a hybrid position in which both systems could play an important role (Pantsar, 2021a; Spelke, 2011). But the fundamental content of the neuronal recycling account would remain the same in connection to the ANS, the OTS and the hybrid accounts: the proto-arithmetical neuronal circuits are recycled for arithmetical purposes, and in the process the recycled regions lose part of their original function.

Neuronal recycling appears to be connected to some kind of *modular* notion of mind, according to which the mind is constructed of domain-specific, specialised modules for different cognitive functions (Fodor, 1983). Perhaps in the most important post-Fodorian development, Carruthers has argued for *massive modularity* of the mind based on an argument that such complex biological systems as humans could only evolve if they are organised in a modular way (Carruthers, 2006). In the case of the development of arithmetical cognition, Butterworth has argued explicitly for the existence of a 'number module' as the basis of arithmetic (Butterworth, 1999).[7]

The modular notion of mind is compatible with the plasticity of the brain, but it suggests limits to this plasticity. Pathological cases, however, show that the human brain is plastic beyond what the modularity account implies. As recounted in Dehaene (2020), children with almost an entire brain hemisphere missing can develop largely 'normal' cognitive abilities. In such cases, the remaining hemisphere accommodates the other cognitive functions. Sur et al. (1988) tested this in ferrets by effectively turning the auditory cortex into a visual cortex. Such findings suggest that the plasticity of the brain can be remarkably high. Indeed, is it high enough to refute the modular notion of mind or the neuronal recycling hypothesis? While such extreme cases of neural plasticity do seem to be evidence against modularity, this is not necessarily the case, since the massive plasticity and re-organisation of the brain only seems to take place in

[7] For more on the issue of modularity, see Fabry (2020).

pathological cases. In phenotypical individuals in the same culture, neuronal circuits standardly develop in similar ways and there is a close resemblance in the structural and functional organisation of their brains (Dehaene, 2020).

Nevertheless, it has been argued that the neuronal recycling hypothesis is potentially too limiting and does not fully capture the degree of plasticity of the brain, nor is it supported by empirical evidence (Fabry, 2020). Michael Anderson has proposed the notion of *neural reuse* as a better organisational principle (Anderson, 2010, 2015). Recently, neural reuse has been argued as the better fit for the enculturation framework, including in the case of arithmetical cognition (Fabry, 2020; Jones, 2020). According to the neural reuse principle, neural circuits do not generally perform specific cognitive or behavioural functions. Instead, neural systems show flexibility and redundancy in how they function. Cognitive and behavioural functions can thus employ and re-deploy neural resources from many neural circuits across different brain areas, and any kind of modular notion of mind is misplaced. Instead, the neural reuse theory allows for quick changes in how the brain organises itself, consistent with many data (Anderson, 2015, 2016).[8]

However, neural reuse should not be confused with 'blank-slate' theories of the mind that argue against innate biases in the way the human brain is structurally and functionally organised. Instead, neural reuse includes the notion that brain regions have *functional biases*, dispositions that make them more suitable for some cognitive or behavioural functions than others. Thus, neural reuse can also account for the fact that, in neurotypical individuals within a culture, cognitive capacities are standardly associated with the same regions of the brain, while also allowing a more flexible organisation in connecting different regions in association with different cognitive tasks.

Max Jones (2020) has used the functional flexibility of the ANS to argue for neural reuse as the organisational principle of the brain involved in enculturation. The ANS can play different roles, ranging from pure estimation tasks to different ways of approximating the size of symbolic numerals (e.g., in terms of estimating the numerosity of the digits or the

[8] Research on non-human primates shows that neural reuse happens on the level of single neurons. In addition, experiments with animals with simple neuronal systems (e.g., the soil-dwelling roundworm *Caenorhabditis elegans*) reveal that the synaptic structure can be largely identical between individuals, but there can be vast functional differences in the particular neurons, up to the point that they are associated with completely opposite tasks (Anderson, 2015; Cisek & Kalaska, 2010; Varshney et al., 2011).

number represented by the numeral). How this happens depends on the particular task which results in differences in the way the ANS is connected to other neural resources. According to Jones, the more general principle of neural reuse can explain this better than neuronal recycling. In a similar way, Fabry (2020) has argued that Menary's notion of learning driven plasticity, which is based on neuronal recycling, would be characterised better in terms of neural reuse. Among other things, Fabry argues that neural reuse is a better fit with the way bodily interactions with the environment shape our cognitive development. Penner-Wilger and Anderson (2013) have argued that the well-known relation between finger gnosis and mathematical ability (Noël, 2005) can be best explained through re-deployment of neural circuits, since part of the same neural circuit is employed. This, however, is only one of the neural circuits that is associated with numerical cognition. This kind of multiple use of the brain region as parts of different neural circuits associated with different cognitive processes appears to be a better fit with neural reuse theory (Fabry, 2020).

When it comes to mathematical cognition, relevant data for the topic of neural reuse have also emerged from fMRI scans of blind mathematicians. It has been established that advanced mathematical reasoning activates bilateral areas of the brain, including the intraparietal, inferior temporal and dorsal prefrontal cortex (Amalric & Dehaene, 2016). Amalric et al. (2018) reported a study of blind mathematicians which showed that they recruit the same brain regions as sighted mathematicians, suggesting that advanced mathematical reasoning is not dependent on visual experience. Interestingly, the fMRI scan also showed activation of the occipital cortex, which in the sighted is used for the visual processing of images on the retina. Indeed, it has been reported that in (auditorily presented) mental arithmetic tasks the only difference between sighted and congenitally blind individuals is that the latter showed additional activity in the occipital cortex (Kanjlia et al., 2016). These data suggest that the visual cortex is partly reused in the congenitally blind for numerical and other mathematical tasks, or as Kanjlia and colleagues put it, 'In blindness, [the frontoparietal number network] colonizes parts of deafferented visual cortex. These results suggest that human cortex is highly functionally flexible early in life' (Kanjlia et al., 2016, p. 11172).

These kinds of data could be seen as evidence for the neural reuse account instead of the neuronal recycling account. It would thus be interesting to know what Dehaene, the main proponent of neuronal recycling, would make of them. Unfortunately, however, in his latest book, Dehaene (2020) does not explicitly discuss neural reuse. However, he does discuss the data on blind mathematicians in terms of 'extreme proof of brain plasticity' (Dehaene, 2020, pp. 129–130), which I understand to be similar to plasticity guided by the principle of neural reuse. In any case, Dehaene sees the data on blind mathematicians as evidence of neuronal recycling, and not as extreme proof of brain plasticity. In support for this, he points to research showing that the visual cortex in the blind seems to mostly maintain the typical connectivity and neural maps (Bock et al., 2015). One interesting case of this is that the area of the brain recruited for written words, the left occipitotemporal sulcus, is the same in sighted individuals as for blind people reading in Braille (Bouhali et al., 2014).

While I agree with Dehaene's assessment that the data on blind mathematicians should not be seen as evidence for the kind of extreme 'blank-slate' plasticity of the brain that overcomes all innate connectivity biases, the data seem to be evidence for neural reuse rather than neuronal recycling. In blind mathematicians, an area of the brain, that is, the occipital cortex, is recruited for mathematical purposes in a way that differs from sighted mathematicians. This suggests that the extent of neural reuse goes beyond recycling particular neuronal circuits. If cognitive processing can include employing neural resources from an area of the brain previously not associated with some cognitive function, this is potentially important evidence toward establishing wider reuse rather than the recycling of particular brain regions.

More empirical research is needed before any conclusive assessment can be made, but I agree with Fabry and Jones that currently there are good reasons to prefer the notion of neural reuse as the organisational principle behind enculturation. Neuronal recycling and the re-deployment of neural circuits can then be seen as one specification of neural reuse, or perhaps it needs to be understood as an altogether different principle. In any case, I believe that the evidence strongly implies that we should avoid commitment to a modular theory of mind when it comes to the development of arithmetical cognition. While it is true that there is very little inter-individual variation in the activation of brain regions in conducting arithmetical tasks, this should not be seen as evidence of a 'number module'. This phenomenon can be explained with a weaker basic assumption: that there is a functional bias to employ parts of the parietal, frontal and temporal lobes, in particular the intraparietal sulci, in the acquisition of number concepts and arithmetical abilities (Nieder & Dehaene, 2009). This functional bias is explained by evolutionarily developed proto-arithmetical abilities, which are applied in acquiring arithmetical ability. Aside from studies on human agents in proto-arithmetical and arithmetical tasks (Nieder & Dehaene, 2009; Piazza et al., 2007), the existence of such functional bias gets support from the similar activation patterns in non-human primate brains in tasks involving numerosities (Cantlon et al., 2016; Nieder, 2016).

Importantly, this functional bias endures throughout the acquisition of mathematical expertise. One often presented idea in the literature has been that mathematical cognition is ultimately an abstraction based on linguistic operations (see, e.g., Chomsky, 2006). This would go against the present account in which the development of arithmetical cognition is seen as a process of enculturation based on proto-arithmetical abilities. Given that proto-arithmetical abilities are non-linguistic, this account is incompatible with that proposed by Chomsky. In the account I have developed, proto-arithmetical abilities are not abandoned when linguistic capacities with numerosities are developed. Instead, arithmetical – and other mathematical – cognition is seen as a phenomenon in which different cognitive capacities are included, only some of which are linguistic.

Ultimately this is a matter to be decided by empirical evidence, and the data thus far strongly suggests that Chomsky is wrong. Indeed, an fMRI study reported by Amalric and Dehaene (2016) shows that, in professional mathematicians, a particular brain network consisting of bilateral frontal, intraparietal and ventrolateral temporal regions activates in processing mathematical statements. This network overlapped with the network consisting of the bilateral prefrontal cortex, intraparietal sulcus and inferior temporal cortex, which is the network associated with proto-arithmetical abilities (Nieder, 2019, p. 232). It is crucial to note that the professional mathematicians did *not* recruit brain areas associated with language. This is strong evidence that mathematical thinking is not simply a special form of linguistic processing, which is consistent with the present account.

3.5 Empirical Predictions

Although one concerns an ontogenetic and the other a phylogenetic process, the enculturation account bears some resemblance to the notion of 'exaptation' in evolutionary biology. Exaptation – in contrast to adaptation – refers to features of biological organisms that enhance fitness but were not developed by natural selection for that purpose (Dehaene, 2005; Gould & Vrba, 1982). Enculturation, on the other hand, makes it possible to reuse (or recycle) neural resources developed by natural selection for other purposes. While exaptations occur on a long phylogenetic timescale of a species, processes of enculturation happen on the timescale of individual ontogeny. An enculturation process can take place in years, months or even days, and it is possible by using evolutionarily developed features for new purposes. Neural reuse and neuronal recycling are both accounts aimed to explain how this is possible. But that such cultural learning *is* possible is beyond any doubt. The question is whether the present enculturation framework captures the phenomenon accurately and in all its richness.

In presenting his neuronal recycling hypothesis, Dehaene (2005) formulated three empirical predictions resulting from the hypothesis. First, our genetically determined cognitive architecture should limit what can be learned culturally. Second, the difficulty of learning a cultural practice or skill should depend on the distance between the initial function of the neuronal circuit and the new one. Third, cultural learning can reduce the cortical resources available for earlier abilities (Dehaene, 2005). I believe that these three predictions can all – *mutatis mutandis* – be extended to concern the enculturation account. In addition, I propose two other predictions. First of these is that we should expect increasing inter-cultural differences when the culturally developed abilities are increasingly distant from the evolutionarily developed abilities whose neural resources they reuse. This is to be expected because the further the culturally developed abilities are from the original evolutionarily developed abilities, the further away their neuronal realisation is likely to be from the functional bias of the brain, which has evolved for the purpose of the original ability. The second new prediction is that we should expect there to be culturally developed abilities that are more difficult to acquire later in ontogeny, due to cortical resources already having been used for earlier abilities. While this may sound like an expectation specific to neuronal recycling, I believe that applies to the principle of neural reuse as well. Also, in neural reuse, cortical resources are increasingly used in the progress of ontogeny, making it more difficult to reuse areas of the brain that are already used for other functions.

To conclude this chapter, let us take a look at the five predictions in the context of mathematical (in particular arithmetical) cognition and briefly discuss what kind of empirical data could corroborate them, and what kind of data would conflict with them.

Prediction 1: Our proto-arithmetical abilities should limit what kind of mathematical abilities can be learned culturally

Clearly there are limits to what kind of mathematical abilities can be learned. For example, for most people, the mental arithmetic involved in multiplying multi-digit numbers becomes quickly impossible. However, it is not clear that such limits are due to our proto-arithmetical abilities. Instead, it seems that working memory constraints and complexity of the calculations are more likely to be the limiting factors.[9] While these may be connected to proto-arithmetical abilities, it is not obvious that in such cases proto-arithmetical abilities limit cultural learning. However, there are other ways in which proto-arithmetical abilities may limit mathematical learning. For example, they may explain why natural numbers are so much easier to learn than other kinds of number systems. If there is a cognitive bias toward linear natural numbers (due to the OTS), that could explain why grasping logarithms and exponentiation is so difficult for many students (see, e.g., Aziz et al., 2017). There are many other ways in which proto-arithmetical abilities may limit cultural learning,[10] ranging from numeral systems (e.g., the difficulty for humans to process a binary system) to the kind of cognitive tools we use (e.g., we can observe only a limited amount of objects in parallel).

Empirical research on such issues can reveal limits that subjects without evolutionarily developed abilities with numerosities, for example, artificial intelligences, may not have. Some such limits have already been established, as reviewed in Section 1.2. Even after acquiring abilities with symbolic numbers, we cannot help reverting to the ANS in number comparison tasks. This way, our completion of the tasks is *computationally suboptimal*, which is already a limit in our arithmetical ability due to the proto-arithmetical abilities, thus corroborating the enculturation theory (Pantsar, 2019a). More support for Prediction 1 is received from data on cases in which arithmetic learning is problematic. Developmental dyscalculia is a learning disorder that causes difficulties in the acquisition of number concepts and arithmetical knowledge and skills. One of the main hypotheses for explaining developmental dyscalculia is the *core deficit hypothesis*, according to which developmental dyscalculia is predicted by impairment in proto-arithmetical acuity with numerosities (Halberda et al., 2008; Knops, 2020; Piazza et al., 2010). Such results imply that

[9] Indeed, working memory level has been established as one of the best cognitive predictors of later success in arithmetic and more generally in mathematics (Geary, 2011).

[10] What exactly counts as 'cultural learning' turns out be a difficult question, which I will tackle in detail in Section 6.2. For now, it is enough to understand it as the kind of learning that is only possible in cultural contexts. The circularity of this characterisation is glaring, but the notion of cultural learning I endorse in Section 6.2 gets rid of it.

problems with proto-arithmetical abilities result in limits in what kind of mathematical contents can be learned in a cultural setting.[11]

Prediction 2: The difficulty of learning a mathematical practice or skill should depend on the distance between the initial function of the proto-mathematical neuronal circuits and the new ones

This prediction states that the further away we go from the proto-mathematical abilities, the more difficult we should expect the learning process to become. Focusing on proto-arithmetical abilities, this means that we should expect abilities with other number systems to be more difficult to learn than arithmetical abilities. Assuming this to be the case is the entire basis of our educational systems when it comes to mathematics. Children are first taught basic arithmetic and, in carefully designed stages, they are introduced to negative numbers, rational numbers, real numbers, and so on. Of course, this could simply be due to an objectively increasing difficulty in mathematics as we introduce new number systems. However, there is also evidence that seemingly equally simple mathematical theories are not equally difficult to learn. Most notably, in US grade schools in the 1950s and 1960s there was the 'new math' innovation that aimed to increase children's understanding of mathematics. Instead of starting from basic arithmetic, children were taught set theory as the basis of mathematics. This led to many students having problems acquiring arithmetical skills (Kline, 1973). There may be multiple reasons for this, but one plausible reason was that arithmetic can be taught as a more natural continuation of proto-arithmetical abilities than set theory. Perhaps there could be also 'proto-set-theoretical' cognition, in which case the failure of the new math movement would likely lie in other factors, but as of now the empirical data supporting the existence of proto-arithmetical abilities is much stronger. Further research is needed, but the ease with which most children in arithmetical cultures learn arithmetic gives support to Prediction 2.

Prediction 3: Cultural learning of mathematical abilities can reduce the cortical resources available for proto-mathematical abilities

The basis of this prediction is that, if neural resources used for proto-mathematical resources are reused for mathematical purposes, we should

[11] The other main hypothesis, the *access deficit hypothesis*, states that developmental dyscalculia is caused by a problem in accessing the proto-arithmetical abilities when learning numerical symbols, and not by a deficit in the proto-arithmetical acuity itself. If the access deficit hypothesis turned out to be correct, developmental dyscalculia would not provide support for Prediction 1.

expect there to be less resources remaining for the former. As we have seen, even after learning symbolic arithmetic, we do not lose our proto-arithmetical abilities. In fact, the ANS-based estimation ability improves throughout early ontogeny, only reaching full acuity in adulthood (Halberda et al., 2008; Piazza et al., 2010). Thus, it does not appear that learning arithmetical abilities causes a decrease of proto-arithmetical abilities. However, this does not mean that there is no difference in the cortical resources available for proto-arithmetical abilities. It could be the case that increased proto-arithmetical acuity is achieved while applying less cortical resources for it than previously. This prediction is as of now, to the best of my knowledge, not strongly corroborated by evidence. However, given increased activity in areas of the brain associated with proto-arithmetical abilities – mainly in the parietal and frontal lobes – also in association with arithmetical abilities, it is at least plausible that less cortical resources are available for proto-arithmetical abilities. This would most obviously be consistent with the neuronal recycling hypothesis, but I believe it is also applicable to the neural reuse account. Due to the functional biases of brain regions, it is plausible that the neural reuse process employs cortical resources in particular regions, leaving less resources for their original functions.

Prediction 4: We should expect increasing inter-cultural differences in mathematical abilities when they are increasingly distant from proto-arithmetical abilities

This topic will be treated in detail in Part II of this book, but for now it suffices to say that this prediction is strongly corroborated by anthropological data. While arithmetic has been developed several times independently by cultures (e.g., Mesopotamians, Mayans, Chinese), there have been important cultural differences (Ifrah, 1998). The notion of mathematical proof, for example, seems to have emerged in the Mesopotamian and Chinese cultures, whereas in the Mayan culture – to the best of our knowledge – arithmetic was limited more to calculations (Chemla, 2015; Ifrah, 1998; Pantsar, 2019b). Generally, the further we go from proto-arithmetical abilities, the more variation there seems to be among cultures, also in terms of numerical abilities (Everett, 2017). The most extreme cases of this are cultures that do not possess numeral words or number concepts, such as the Munduruku and the Pirahã (Gordon, 2004; Pica et al., 2004).

Prediction 5: We should expect there to be mathematical abilities that are more difficult to acquire later in ontogeny, due to cortical resources already having been used for earlier abilities

This last prediction is based on the generally accepted idea that there are 'sensitive periods' in ontogeny, meaning that it is easier for an individual to learn certain abilities during a particular period in their development. In the case of language, for example, children show lower learning results already after the first year of their life (Dehaene-Lambertz & Spelke, 2015). In the case of mathematics, the sensitive period appears to start in pre-school age (Clark et al., 2010). While adults do not lose in a significant way the ability to learn mathematics, adult learning of elementary mathematics poses important challenges that are (partly) different from children's education (FitzSimons & Godden, 2000).[12] This may be indicative of missing the sensitive period for learning mathematics, connected to using cortical resources for other abilities. Again, the most extreme cases come from cultures where individuals have acquired no mathematical knowledge before adulthood, such as the Munduruku and the Pirahã. Anthropological reports imply vast difficulties in grasping basic mathematical concepts, like that of one-to-one correspondence (Everett & Madora, 2012).

As with Prediction 3, it might seem that this prediction is only plausible if we accept the neuronal recycling hypothesis. However, also in this case the functional biases that brain regions have make the prediction applicable to the neural reuse account as well. If a child does not learn arithmetic during their sensitive period, it can be the case that the areas of brain with functional bias for arithmetic are employed more firmly for other purposes. In this case, according to the neural reuse account, either the areas with the functional bias need to be redeployed for arithmetic, or other cortical resources need to be employed in learning arithmetic. Importantly, these other cortical resources would include areas of the brain that do not have functional bias for arithmetic. In both cases, it is at least plausible that the process of neural reuse manifests itself as lower results in learning arithmetic, as predicted by the idea of sensitive periods.

3.6 Summary

In this chapter, I focused on the cultural influences in the ontogenetic acquisition of number concepts and arithmetical knowledge and skills. Using the notion of enculturation as specified by Menary, I provided an account of how both cultural and evolutionarily developed biological

[12] For example, adult learners of basic mathematics often require a more direct connection to actual life situations.

aspects influence the acquisition of numbers concepts and arithmetic in the individual. I discussed the neuronal mechanism that enables enculturation processes, reviewing literature that suggests Anderson's notion of *neural reuse* to be a better fit than Dehaene's notion of *neuronal recycling*. Finally, I presented five empirical predictions resulting from accepting the enculturation account of arithmetic in ontogeny.

PART II

Phylogeny and History

CHAPTER 4

The Phylogeny and Cultural History of Number Concepts

In Part I of this book, the focus was firmly on the individual. I argued that the enculturation account provides a feasible theoretical framework for studying the development of arithmetical knowledge and skills in individual ontogeny. Enculturation, however, is only possible when the individuals are members of cultural communities. Learning in a culturally influenced context is shaped by culturally developed contents, artefacts and practices, among other factors. For processes of enculturation to take place, an individual needs to interact constantly with members of the culture, getting access to the knowledge and skills shared by them. Therefore, in addition to the individual, we must extend our approach to include the way members of cultural communities can develop those skills and knowledge.

When it comes to the question of arithmetic, these questions are particularly interesting. As we will see in this second part of the book, arithmetic has a distinctive cultural history. Unlike language, for example, arithmetic is not a universal ability. While *proto-arithmetical* abilities are universally shared by humans, there are many cultures, such as the Pirahã and the Munduruku, that have not developed their ability with numerosities substantially beyond early proto-arithmetical origins (Gordon, 2004; Pica et al., 2004). But also in cases where cultures have developed arithmetic, there have been important differences in the trajectories (Ifrah, 1998). In terms of cultural history, arithmetic is not a uniform discipline that follows a single cultural tradition. It has been developed independently several times, including by the Chinese, Mesopotamians and Mayans, every time taking distinctive forms and serving different purposes. However, the development of arithmetic has also every time converged in important ways. The results of basic arithmetical operations (addition, multiplication, subtraction, division) for finite numbers do not vary from one culture to another. When the Mayans could calculate results of operations up to millions, they got the same results as we do in our

Mesopotamian tradition (Ifrah, 1998). Arithmetic is not universal, and even in cultures that have developed arithmetic, it does not always have the same characteristics. But an important part of arithmetic *is* shared between cultures that could not have feasibly had any direct contact between each other. Explaining this convergence of arithmetical knowledge across cultures is one of the key challenges for the epistemology of arithmetic.

In order to be able to analyse why different cultures develop arithmetic in convergent ways, we need to generally understand how cultures can generate and disseminate knowledge. We will turn to this question in Chapter 6. First, however, it will be important to understand how arithmetic has developed culturally, starting from number concepts.[1] Before we continue, however, I want to make one terminological clarification. In addition to cultural history, I write here about *phylogeny* of numbers, but 'phylogeny' as a term is traditionally associated with biological evolution. Thus, the use of this term may seem to contradict my position that numbers are a cultural development. However, numbers are not *only* a cultural development, given their origins in biologically evolved proto-arithmetical abilities. Therefore, I sometimes write about the development of numbers also in terms of their phylogeny. But when the topic is specifically the cultural development of natural numbers based on proto-arithmetical abilities, I will write about the *cultural history* of numbers.

4.1 Can There Be Numerals before Number Concepts?

The enculturated account of number concept acquisition presented in Chapters 2 and 3 can only make sense if members of cultural communities possess number concepts that they are able to share. This starts already from the first stage of the bootstrapping process, that is, acquiring the placeholder list of numeral words. Such lists are cultural constructs and not universally present. This prompts the question: Where do numeral word lists come from? If the enculturated bootstrapping account is correct, we need numeral words in order to acquire number concepts. But how can numeral words develop if there are no number concepts that they refer to? This question has been posed in many different ways in the literature, but here I follow the formulation by Jean-Charles Pelland (2018, 2020).

[1] In what follows, for reasons of readability, I will often use the word 'culture' to refer both to a cultural community or a cultural group and an inter-generational notion of culture. Thus, when I write about, for example, 'the Mayan culture', it can mean both the culture that the Maya developed from 2000 BC to AD 1500 and the cultural community of Mayan people at a particular time. I trust that the context will make clear in which sense the word 'culture' is used.

In general, the question is how number concepts could emerge in cultures where there previously were no number concepts. Specific to the enculturation account, the question is how people could initially acquire number concepts when their acquisition requires other members of the culture to possess them. The enculturated bootstrapping account seems to face this problem in a particularly serious manner, given its dependence on numeral words for the process of number concept acquisition.

In the case of many cultures, numeral words of course come from interactions with other cultures (see, e.g., Everett, 2017). Just like many other cultural innovations, numeral words can be subject to inter-cultural transmission. Most obviously this has been the case with numeral symbol systems. The Indo-Arabic numeral symbol system is currently used worldwide (although not everywhere and not always exclusively). However, it used to be only one of many numeral symbol systems and took a long time to establish itself, for example, in Europe (Ifrah, 1998). Linguistic evidence suggests that numeral words have also been transmitted from one culture to another (Ifrah, 1998). Thus, the question where numeral words come from, in most cases, gets the philosophically uninteresting answer: from other cultures. However, clearly this could not have been the case for all cultures. Some culture or cultures must have developed numeral words on their own. And if they did not already possess number concepts, they must have developed these too. It is this scenario of creating numeral words and/or number concepts from nothing that the challenge formulated by Pelland concerns.

Pelland himself sees the challenge as implying some kind of nativism concerning number concepts: numeral words can only have referents if we possess pre-verbal number concepts that they connect to (see also De Cruz & De Smedt, 2010). However, there is no evidence of such nativist number concepts. As we saw in Chapters 1 and 2, there is only evidence of innate proto-arithmetical abilities which neither requires nor supports the existence of innate number concepts. Number concepts, based on the state-of-the-art evidence, are a cultural construction, and as such Pelland's challenge should not be seen as implying a nativist explanation. The challenge itself, however, is valid. We need to explain how number concepts have emerged in cultures that have previously only possessed proto-arithmetical abilities with numerosities. In order to do that within the enculturated bootstrapping framework, we first need to explain how numeral words could have developed without there being prior number concepts.

This question of the relation between number concepts and numeral words may sound like a 'chicken or egg' type of dilemma, and in a way it is. However, just like the question of whether the chicken or the egg preceded the other, the question concerning the priority of numeral words or number concepts is not some kind of unsolvable puzzle. Unlike the colloquial talk of the chicken or egg dilemma suggests, it is far from a genuine dilemma. By gaining more knowledge about biological evolution, we can explain how new species of fauna develop. Analogously, with improved understanding of cultural evolution, we can explain better how concepts (and words) emerge. Indeed, as I will show in Section 4.4, the best understanding we currently have strongly suggests that – just like the chicken and the egg – number concepts and numeral words co-evolved.

4.2 Material Engagement Theory and Cultural History

In Section 3.1, we saw how under the present approach the ontogenetic development of numerical cognition should be seen as embodied. Finger counting, counting games with physical objects and other aspects of embodied cognition were identified as being key to the enculturated acquisition of number concepts and the development of arithmetical abilities. In this way, this book is located in the '4E cognition' landscape, referring to theories according to which cognition is in some way embodied, embedded, enacted and/or extended (De Bruin et al., 2018). So far, we have been concerned with the ontogenetic aspects of embodied cognition, but I will show that they are closely connected to the phylogeny and cultural history of arithmetic. This is important because there is very little evidence available on the cultural origins of arithmetic, and even the little that we have is often open to several interpretations. Thus, when it comes to early developments in arithmetic, such as the introduction of number concepts, the available historical and anthropological data underdetermines any conclusions. However, the connection to ontogeny can help us formulate plausible hypotheses of the phylogeny and cultural history of number concepts. Given that the genetic make-up of human beings has not significantly changed from the times when arithmetic was developed, observations of modern cultures can help us understand past cultures and their history. Hence, to understand the phylogeny and cultural history of number concepts and arithmetic better, let us first continue for a while with ontogeny, from the perspective of 4E cognition.

When it comes to arithmetical and numerical cognition, the 4E cognition direction has been recently pursued by several philosophers (e.g., Fabry, 2020; Gallagher, 2017; Hutto, 2019; Menary, 2015; Zahidi, 2021; Zahidi & Myin, 2016). However, these approaches are very different from each other. While the work of Menary and Fabry, for example, is positioned in weaker forms of the 4E landscape, that of Zahidi, Myin and Hutto is 'radically enactivist'. According to the enactivist view, as in all 4E accounts, cognition arises through embodied interactions between an organism and its environment (Varela et al., 2017). According to the radical version of enactivism, however, 'basic' cognition does not involve mental representations or mental content (Hutto & Myin, 2013). Representations and content are something that arise only once there are linguistic truth-telling practices in place (Hutto & Myin, 2013; Zahidi & Myin, 2016).

While I am sympathetic to the 4E landscape in general, I see no reason to commit to such radical enactivist views. As mentioned in Section 1.1, there is no convincing argument why the ANS and/or the OTS could not function based on genuine mental representations of numerosity, like those of occupied object files (in the case of OTS) (see Pantsar, 2023c, for details). These should not be confused with linguistic representations, nor should they be thought to imply the existence of innate number concepts. Nevertheless, the enculturation account of number concept acquisition and the development of arithmetical cognition in ontogeny is consistent with there being proto-arithmetical, non-symbolic representations of numerosity, the kind that radical enactivists deny. However, my account is not committed to the existence of such representations either. Perhaps there could be an explanation that does not require numerosity representations at all, as proposed by Zahidi (2021). But as things stand now, there is no reason to prefer radical enactivist accounts of proto-arithmetical cognition over representationalist accounts.

Having said that, there is a lot to learn from the radical enactivist accounts of numerical cognition, like those presented in Zahidi (2021) and Zahidi and Myin (2016, 2018). This is not because they manage to show that proto-arithmetical cognition does not involve representations, but because they are developed in a framework that does not assume the existence of *innate* numerosity representations. There is no compelling argument or evidence why proto-arithmetical abilities could not involve representations, but neither is there any reason why these abilities should be connected to innate representations. If the object file hypothesis, for example, is correct, then the OTS functions based on occupying mental object files. The occupied object files are then feasibly understood as implicitly representing numerosity. Importantly, this does not mean that there is an innate representation of, say, the numerosity three. What it means is that there is an innate cognitive *mechanism* that can provide a means for representing the numerosity three. From a general enactivist point of view, this should be quite acceptable, since the representations are thought to arise only through interactions with the environment (i.e., by applying the object tracking system). In the *radical* enactivist view this is not acceptable, given that all representations are thought to require language. However, the wider principle of such cognitive mechanisms should not be problematic for radical enactivists either. After all, they do accept that there are innate mechanisms that make representations possible. The main difference is that for the radical enactivists this requires linguistic practices, while in my account it does not.

In this manner, I want to detach my account from the radical enactivist position, while in general being open to 4E views concerning cognition. For example, I am sympathetic towards the basic idea of *material engagement* as proposed by Lambros Malafouris (2013). Malafouris opposes the view that cognition is limited to the brain and instead emphasises the embodiment thesis that cognitive processes should be understood in terms of the way our brains, bodies and environments interact. While I remain sceptical about the way Malafouris wants to eradicate representations from cognition, I agree with him that we cannot detach cognitive phenomena from the way they are enacted in material surroundings. When it comes to the development of numerical cognition in ontogeny, the importance of engaging in counting games and finger counting, for example, fits well with this framework.

Importantly for the present topic, in addition to ontogeny, I contend that material engagement theory can also help explain the *phylogeny* and *history* of arithmetical cognition. This topic is divided into two questions. The first question concerns the *biological* evolutionary phylogeny of proto-arithmetical abilities. The second concerns the *cultural* evolutionary history of number concepts and arithmetical abilities. As for the first question, at this point I do not have much more to add. Subitising and estimating are both proto-arithmetical abilities that have developed through processes of biological evolution. As such, their phylogeny should be analysed in terms of processes of natural selection that (mainly) drive biological evolution. We can speculate on the reasons why proto-arithmetical abilities turned out to be evolutionarily advantageous. It is easy to accept, for example, that being able to determine the numerosity of prey animals or predators was an ability that helped with survival. While these are extremely interesting questions, for now it is enough to point out that proto-arithmetical abilities *must* be the product of biological evolution. If they were not, we could not share them with many non-human animal species, nor could we possess them already as infants.

However, the cultural history of number concepts and arithmetical abilities is a whole different question. Since it is distinct from biological evolution, we cannot assume natural selection as a driving factor. However, there must have been *some* way in which number concepts and arithmetical abilities turned out to be advantageous for cultures. We will return to this question soon, but first we need to focus on a more fundamental question. How is the cultural evolution of skills and knowledge possible in the first place? How are abilities and knowledge passed from one generation to another? Indeed, they are not simply passed, they are also further developed by new generations. How does this process take place?

In the rest of this chapter, I will focus on these questions concerning natural number concepts, while in Chapter 5 I analyse the questions in relation to the development of arithmetical skills and knowledge. In Chapter 6, I will then present an integrative theoretical framework based on the phenomenon of *cumulative cultural evolution* in order to make sense of this kind of cultural history of concepts. In all this, my approach is consistent with accounts of embodied cognition, like the material engagement theory. As we will see, forms of material engagement with objects and tools can give us plausible explanations of the different stages in the development of arithmetic, starting from numeral words, numeral symbols and number concepts.

4.3 The Phylogeny and History of Numeral Systems

When it comes to the history of the first systems of numeral words and symbols, the unfortunate fact is that very little is known for sure. The origins of the intellectual development that has led to our numeral systems are irrevocably lost in pre-history. We do not know when the first numeral words emerged and where. In fact, we do not even know whether *Homo sapiens* was the first sub-species of humans that had them. If *Homo neanderthalensis*, as is now generally accepted, had linguistic abilities (see, e.g., Conde-Valverde et al., 2021), we cannot know whether they included numeral words or not. Furthermore, we do not know whether people had a notion of abstract numbers before they had numeral symbols or words. If the account developed in this book is correct, that is unlikely; but it nevertheless is a possibility that cannot be ruled out.

However, that is not to say that we cannot know anything about the early origin of numeral words. Here the material engagement theory can be helpful, as argued by Karenleigh Overmann (2018). While it is impossible to know the kind of numeral words that the first 'numerical' people – whoever they were – used, many human artefacts have remained to give us information of the way numerosities have been represented physically. The earliest of these, according to Overmann, can be found in the Cosquer and Gargas caves in France. On the walls of these caves are hand stencil paintings that depict left hands with extended digits, from one to five. Overmann argues that the hand shapes represent a counting technique from one to five, starting from the extended thumb representing one, and all the digits including the little finger extended representing five (Overmann, 2021a). These cave paintings from the Paleolithic era are about 27,000 years old. In case the hand stencils truly did serve numerical purposes, they could be the oldest (thus far discovered) human artefacts to represent numerosities. Their best-known competitor for that title, the Ishango bone found in Congo, dates to about 20,000 years ago, but even older bones with clearly human-made notches have been found, including a 37,000 year-old piece of baboon fibula, nowadays standardly referred to as the Lebombo bone, with 29 regularly spaced tally marks (Bogoshi et al., 1987). However, it is difficult to know whether the notches represent numerosities – let alone numbers – which makes the claim of Bogoshi et al. that it is the 'oldest mathematical artefact' highly dubious. It may have been used for tallying, that is, counting objects by making notches in the bone, but the notches could have been also, for example, purely ornamental.

In this regard, the Ishango bone (Figure 4.1) is different. It has different sizes of groupings of tally marks in two columns, with both adding up to sixty (Everett, 2017, p. 35). The significance of the markings has been debated, but the groupings and the same, quite large, sum in the two columns suggest that the notches serve some systematic purpose related to numerosities. As Everett points out, one end of the bone has a sharp piece of quartz sticking out of it, which makes it likely that the bone itself was used for engraving. If that is the case, it is possible to hypothesise that the groupings of tally marks served as some kind of reference for the engraver.

Here it is not possible to go into that debate, but while it would be a stretch to call the Ishango bone a mathematical artefact, it also seems unlikely that the notches did not have anything to do with numerosities. A wealth of comparable objects with tallying marks have been found from more recent times and it is very likely that such a tallying system formed an early basis for at least the Roman numeral symbol system (Ifrah, 1998; see also Stjernfelt & Pantsar, 2023).

Figure 4.1 The Ishango bone, the oldest known artefact to show a tallying system for numerosities.
Image: Author.

It is likely that tallying systems, one of which probably was responsible for the notches on the Ishango bone, were an early method of keeping track of quantities. The same could be true of the hand formations present in the Cosquer and Gargas paintings but, regardless of the exact details concerning particular objects, it is well established that hands and other body parts have been used widely for representing quantities. Indeed, they are still used commonly in learning to count and number concept acquisition in children, but also in many parts of the world for both counting and calculations outside educational contexts. A finger-counting method using joints and knuckles emerged in Ancient Egypt and preserved until the Middle Ages in Europe that was able to represent numbers from 1 to 9,999, and the Chinese finger-counting system had an even wider scope (Ifrah, 1998, p. xx). In addition to hands, the counting systems often involve other body parts, and pointing to a particular body part is used to express a particular quantity. Such body part counting systems are often found in non-arithmetical cultures that do not possess numeral symbols (Butterworth, 1999; Everett, 2017). But the most common way of using body parts in numerical contexts seems to be connected to counting. This can explain why most numeral systems have ended up using the base ten, and also why many others used the base of twenty (including both fingers and toes) (Ifrah, 1998, p. xix).[2]

Body parts and tallying, however, are both rather deficient methods of keeping track of quantities. With body parts, for one thing, communication of a quantity requires two or more people to be present in the same place. For tallying, one clear drawback is that the amount of notches quickly becomes too large to easily grasp. Furthermore, moving to arithmetical operations, subtraction is difficult because, once a notch is made, it cannot be removed in a straightforward manner. Such difficulties are not present when quantities are represented by manipulable small objects, like pebbles. Pebbles have been used for calculations (quite literally, since the word *calculus* in Latin means pebble) for at least 5,000 years, and they are responsible for the two most widespread methods of calculation in the pre-electronic calculator times. The abacus is still widely used, mainly in parts of Asia, and it is a direct descendant of using pebbles for representing quantities. Less obviously, however, also the use of written numeral symbols in Western culture most likely has its origin in the manipulation of pebbles, or similar small objects. In around 8,000–4,500 BC, both in Elam (what is now Iran) and Mesopotamia (modern-day Iraq), people started using clay tokens to represent quantities in accounting (Overmann, 2018; Schmandt-Besserat, 1996). Different tokens represented different quantities and the proper array of them was put in a clay case to represent the desired quantity. But what happened afterwards transformed human cultures: to save the trouble of opening the clay case, people started marking the outside of the case with symbols to signal what was inside (Ifrah, 1998, p. xx). From this, the next development was to dismiss the clay tokens inside the case and only use the markings (Nissen et al., 1994).[3]

It is unclear what the exact connection was, but this development happened concurrently with the emergence of the alphabet and writing

[2] Although it raises the question of why some cultures developed a base-12 numeral system and others, like the Sumerians, a base-60 system. It should be noted, however, that the Sumerian/Mesopotamian system was not a pure base-60 system, since it did not have sixty different symbols for the digits. Instead, in terms of its symbols the system was a mix of base-10 and base-6, which makes it a more natural extension of finger counting procedures.

[3] How did such clay markings come to represent numbers in an abstract sense? Valério and Ferrara (2022) have recently challenged the standard story according to which there was a clear development from concrete numerals (used to count particular types of objects) to abstract numerals (used to count any objects) in Mesopotamia (see also Overmann, 2021b). Indeed, it is likely that the historical development was complex and non-linear. Unfortunately, it is not possible to go into the details here.

(Schmandt-Besserat, 1996). Writing transformed human cultures and made the transmission and preservation of information possible in entirely new ways. As pointed out by Everett, the emergence of numerals – both as words and symbols – had a huge influence for cultures also aside from the connection to writing:

> [Numerals] were quite likely foundational to the advent of writing around the world. It is commonly recognised that the scientific revolution, industrialization, and modern medicine were dependent on specific mathematical practices. Millennia prior to the existence of these practices, though, verbal numerals and inscribed numerals helped enact profound changes in how humans subsisted and how they used symbols to convey ideas. (Everett, 2017, p. 238)

The importance of numeral symbols for the development of arithmetic was fundamental, as we will see in Chapter 5. But how did numeral words relate to this development? Unfortunately, there is no record of the prehistorical development of how numeral words were formed and used in the traditions where numeral symbols emerged, like that in Sumer. However, it is possible to see ways in which systems of tallying, body part counting and pebble manipulation may have influenced the emergence of verbal numeral systems. Finger counting, for example, can help establish the linear order of counting, which can then determine the order of numerals (Bender & Beller, 2012). Another possibility is that words for body parts can gradually evolve new meanings as numeral words.

Is there evidence of this happening? Certainly, there exists a lot of circumstantial evidence. Base-10 and base-20 being the most common bases for numeral word systems speaks in favour of the development of numerals being connected to body part counting. In addition, many numeral words in different languages have been traced to body parts (Everett, 2017; Ifrah, 1998). However, the strongest evidence of this transfer of meaning of words towards numerals may come from contemporary cultures which are at different stages of developing numerical languages. In anthropological studies, many cases have been established in which words that were not originally used to signify quantities have acquired numerical purposes. For example, in the Hup language spoken by several cultures in the Amazonia, the word for two, *kəwəg-ʔap*, means 'eye quantity', literally translated.[4] In addition, the word for three, *mɔ̃t-wɨg-ʔap*, is derived from the word for a three-chambered rubber plant seed,

[4] To be more precise, this is one of the variants of the word for two (dos Santos, 2021, p. 9270).

literally 'rubber tree seed quantity' (dos Santos, 2021; Epps, 2006). This suggests that numeral words can emerge from prototypic representations of quantities of natural phenomena. While the historical development of this in the case of Western culture may be impossible to trace, in contemporary (and unfortunately endangered) cultures, this kind of transformation of meaning is taking place all the time (Everett, 2017, p. 238).

As we learn more about the emergence of numeral words, we are likely to gain a better understanding of the influences that are at play in such transformations. However, if we only focus on numeral words (and symbols), we will not get an answer to the question regarding the origins of number concepts. The transfer of meaning explains how numeral words can emerge, but how do they become connected to number concepts? Therefore, the question remains: unless people already possess number concepts, how can words transform their meaning to refer to number concepts? In the cultural history of numeral words and symbols, we still need to ask when *numbers* entered the picture.

4.4 Co-evolution of Numerals and Number Concepts

When trying to answer the question of how number concepts have emerged, the conceptual distinctions made in the Introduction prove to be priceless. Since proto-arithmetical abilities concern numerosities and not numbers, the question of the emergence of number concepts should be understood in a cultural context. In a particular cultural context, it is then most likely that the emergence of number concepts is closely linked to the emergence of numeral words and symbols. Rejecting the nativist view concerning number concepts, the crucial question is not how numeral words and symbols are connected to number concepts. It is how the development of numeral words and symbols are connected to the *development* of number concepts.

The account of material engagement and the development of numeral words and symbols can help us explain how that process takes place. As the existence of anumeric cultures like Pirahã and Munduruku shows, not all cultures have numeral words. The three numeral-like Pirahã words, for example, translate roughly to 'one' (*hói*), 'two' (*hoí*) and 'many' (*báagiso*), but even the use of these words is not consistent in the same sense as our numeral words (Everett, 2017; Gordon, 2004). Moreover, the Pirahã do not seem to have the notion of one-to-one correspondence (for collections larger than three) that is central to acquiring counting processes (Everett & Madora, 2012; Frank et al., 2008; Gordon, 2004). This is evidence that even the ability to do transitive counting – through establishing one-to-one correspondence between numeral words and objects – is a cultural development that is dependent on the existence of numeral words.

Seen from this context, the key question is what happens when, for example, the Hup speaker grasps that the word for the rubber plant seed also refers to the numerosity three? The answer must be found in similar cognitive processes to those discussed in Chapter 2 with regard to bootstrapping. When a child hears the word only in connection to the rubber plant seed, they are unlikely to grasp its numeral meaning. However, when the word is connected to something like counting games, it becomes clear that the meaning of the word captures something other than the seed. It is this cultural practice of connecting a word originally referring to a natural object to new cultural practices that can explain the transfer of meaning. Since the Hup-speaking children can have an OTS-based representation of the three chambers of the seed, they can connect the word with the numerosity three, thus developing the number concept THREE.

As argued by César dos Santos (2021), this is in line with Decock's idea that anumeric cultures use so-called canonical collections of permanent objects as the standard for comparison via one-to-one correspondence (Decock, 2008, p. 464). The problem with Decock's idea is that the evidence points to establishing one-to-one correspondence as already being a culturally developed ability, given that the Pirahã, for example, do not possess it (Everett & Madora, 2012; Frank et al., 2008). However, as in the bootstrapping process, the correspondence can also be established implicitly through the OTS. That way, the representation of a canonical collection – like the rubber plant seed – can be systematically connected with a particular amount of object files being occupied, making it possible to develop number concepts. Dos Santos (2021) provides many more examples in the Nadahup languages (of which Hup is one) where natural phenomena, such as animal footprints, function as such canonical collections, the words for which have developed into numeral words.

The Hup language is particularly interesting for us since in it we appear to capture a language that is at an intermediate stage of the development of numeral words. As reported by Epps (2006, pp. 270–271), the Hup language appears to have exact numerals only up to five, that is, only those numeral words are 'associated with consolidated cardinal values' (dos Santos, 2021, p. 9278). For larger numerosities, gestures and systems of tallying are required, and the numeral words do not have exact meanings. Could the original numeral words that the English words are based on have developed similarly? It is impossible to know, but this explanation seems feasible. The exact phylogeny and history may vary, but it seems that any plausible hypothesis concerning the development of numeral words should be sensitive to the influences of canonical physical collections, body part counting and tallying procedures.

However, there remains the question of how this could have happened for the first time. How did the word for the rubber plant seed, for example, first become connected to the number concept THREE? Again, it is impossible to trace the origins, but the most likely proposed explanation is that numeral words and number concepts *co-evolved* (dos Santos, 2021; Wiese, 2007). There was no number concept THREE beforehand, so it is unlikely that the word suddenly started to carry a numerical meaning. Instead, it is more likely that, by connecting the word for the seed as a canonical collection with a body part counting procedure, the word for the rubber tree seed started to serve another purpose. As words become to play a part in counting word lists, their original meaning can gradually disappear (in that context). As described by Wiese:

> The use of fingers (and other body parts) as tallies can lead to the emergence of a stable conventional order and hence give rise to a second stage in number development: when fingers are used to represent elements of another set, they tend to be singled out successively, following the natural order of fingers on each hand. (Wiese, 2007, p. 766)[5]

This emerging order of counting words can then be used for finger counting processes in which the final word of the recited counting list has a special place. The last word in a list is remembered more easily than the others, which, according to Wiese, gives it a special status. When counting words are used in the same order in association with a given cardinality, the same word will always come last and, thus, get this special status, which supports the indexical link between that word and a certain cardinality (Wiese, 2007, p. 767). This way, words earlier connected with other meanings can become associated with exact numerosities. Dos Santos (2021) explains this in terms of the notion of *re-semantification* (Dutilh Novaes, 2012): words that have lost their original meanings in a 'de-semantification process' and become parts of the counting list now acquire new meanings as referring to exact cardinalities.

This is how the co-evolution of number concepts and numeral words can take place. In cultures where there were previously no number

[5] This quotation is also used by dos Santos (2021, p. 9277), whose influence for this part of my argumentation I acknowledge with great gratitude.

concepts, due to the shared connection to numerosities through proto-arithmetical abilities, people can share an implicit understanding of a numerosity associated with a canonical object. Then the word for this object can be introduced to the body part counting (or tallying) process in which it develops a new use as a part of the counting practice. Gradually, the word loses its original meaning (or at least the earlier exclusive connection to the original meaning; it can still serve a double meaning) as it is increasingly associated with the counting practice. As the counting practice gets more firmly established within the culture, the word acquires a new meaning associated with a particular numerosity in the re-semantification process. As a result of this development, the word is now consistently associated with a particular stage in the counting process applied to discrete objects, and because of this it is used consistently to refer to a particular cardinality of objects. This practice is shared among the members of the culture, so the word has developed a specific meaning to refer to a particular exact cardinality, that is, it refers to a particular culturally shared *number concept*.

In such a development, there is no longer any mystery about the challenge posed by Pelland, that is, where the number concepts came from. They co-evolved alongside the numeral words. Where there was earlier only vague cognitive processing of numerosities, the evolving practice of numeral words used in counting lists also enabled the development of increasingly clear and precise concepts for processing and communicating numerosities. As the counting practices with numeral words develop increasingly exact and consistent forms, also the concepts for processing and communicating numerosities become more exact. After sufficient development, the numeral words and the number concepts follow a standard ordered sequence. When the numeral words (or numeral symbols; the development could happen in a different order) develop a recursive character, it becomes clear that there is no end to them, hence there can be no end to number concepts.[6] At this point it can be said that the culture has developed natural number concepts, which its members can then acquire through the bootstrapping process. There is no chicken and egg problem: just like the bird gradually evolved into a chicken as the eggs also gradually changed, the cognitive processing of numerosities gradually developed into natural number concepts as the numeral word lists and

[6] In modern terms, we would say that the number concepts follow the *omega-sequence*, that is, the ordered sequence 1, 2, 3, 4, . . . This also explains why the number systems do not contain loops, as discussed in Section 2.5.

counting practices developed. Therefore, the challenge posed by Pelland (2018, 2020) and others is met: we can explain how number concepts can emerge when there previously were none.

4.5 Summary

In this chapter, I moved the focus from ontogeny (the level of the individual) to phylogeny and cultural history (the level of communities). Starting from the question which came first, numerals or number concepts, I reviewed literature (e.g., Overmann) applying the material engagement theory of Malafouris, according to which numeral systems developed as the result of interactions with our environment. Following the work of dos Santos and Wiese, I then demonstrated how the most plausible explanation of the development of number concepts is that they co-evolved alongside numeral word systems.

CHAPTER 5

The Development of Arithmetic

In Chapter 4, I have presented an outline of an account that explains how cultures developed natural number concepts. However, number concepts – while certainly already by themselves useful for counting and as such for, for example, very basic bookkeeping – are obviously not yet sufficient for arithmetic. For arithmetic we need numbers, but we also need operations for numbers. Yet even this would be an extremely limited approach to arithmetic. Arithmetic is a cultural activity that has, as we will see, taken many different forms in history. The development of arithmetic within a culture cannot be detached from its wider role in the culture. This includes arithmetical education and arithmetical applications, but also the way arithmetic is positioned in the general cultural scientific and/or intellectual landscape. In this chapter, my focus is on the cultural development of arithmetical skills and arithmetical knowledge in this wider cultural setting.

As in the case of number concept acquisition, I approach the topic here through exploring the connection between ontogeny on the one hand and the phylogeny and cultural history on the other hand. The main reason for this is the dearth of material we have concerning the historical development of arithmetical abilities. As far as establishing the phylogeny and cultural history of, say, addition goes, little can be said with any degree of confidence. However, I believe it is plausible that the way addition is generally learned within a culture in ontogeny is indicative of the way it has developed culturally into the arithmetical concept we are familiar with. For this reason, I ask the reader to bear with me if it feels like we are still mostly dealing with ontogeny. In one way, we are. But it is important to present some more material about the ontogenetic development of arithmetical abilities to establish how they can plausibly mirror the phylogenetic and cultural development of arithmetic, starting from basic arithmetic of simple operations and ranging to formal proofs.[1]

5.1 Basic Arithmetic

The fundamental assumption I make, supported by the state of the art in empirical research, is that no arithmetical skills are included in proto-arithmetical abilities. As detailed in Chapter 1, I see the talk of infant arithmetic and (non-human) animal arithmetic as highly misleading. While arithmetic can be used to explain infant and animal behaviour, there is no reason to think that any of the reported infant or non-human animal behaviour displays arithmetical skills. Arithmetic, starting from the basic addition of small numbers, is entirely a culturally developed ability. The reason for the misunderstanding – aside from applying confusing terminology – is that, for very small numbers, animal and infant behaviour may be largely indistinguishable from displaying proper arithmetic ability. From a purely behavioural perspective, the infant behaviour in being surprised by two dolls being put behind a screen and there only being one doll when the screen is lifted is also what we would expect if the infant calculated $1 + 1 = 1$ and was surprised by the flawed arithmetic. But as we have seen, from a cognitive perspective, these should be treated as two quite different procedures.

For this reason, just as we need to distinguish between proto-arithmetic and arithmetic, we should also distinguish between *proto-addition* and arithmetical addition, and consequently also *proto-multiplication* and arithmetical multiplication (and similarly for subtraction and division). Arithmetical operations are general procedures that can (in principle) be used for any natural numbers m and n. As m and n become larger, calculating the result may become exceedingly difficult to do without the help of cognitive tools such as pen and paper, an abacus or an electronic calculator. With multiplication, this is already the case for most people with two two-digit numbers. The most likely explanation for this is found in the limits of working memory capacities (Imbo et al., 2007).

However, even for sums or products that we are unable to calculate by mental arithmetic (i.e., without applying any external cognitive tools), arithmetically skilful people know that the same procedure works for small numbers as for large numbers. This is not the case with proto-addition. In the cases reported of non-human animals like chimpanzees (Boysen &

[1] For a similar account that distinguishes between the development of basic arithmetic and advanced arithmetic, see Ferreirós (2016).

Berntson, 1989) and grey parrots (Pepperberg, 2006, 2012), there are clear limits to the range in which the animals can proto-add. The chimpanzee studies by Boysen and Berntson had this limit at four, whereas Pepperberg's grey parrot was able to do proto-sums up to six. This difference may not seem huge, but it is in fact highly important: while four is still within the subitising range, six goes beyond it. Thus, while it is possible that the observed chimpanzee proto-addition ability is simply based on the occupation of object files, applying the OTS, something else must be involved in the reported parrot behaviour.

The interesting thing about the grey parrot reported by Irene Pepperberg (2012) was that he was also able to recognise Indo-Arabic numeral symbols and connect them to quantities. In addition, he showed understanding of the numeral symbols forming a progression. What was happening in the apparent proto-addition, according to Pepperberg, was possibly that the parrot formed a numerical representation of the first addend (which were all pieces of food) and then 'counted up' the next addend. This is conjecture, but it is important to note that something other than OTS was at play, since the parrot could proto-add three numerosities up to six (Pepperberg, 2012, p. 715). Pepperberg's grey parrot became something of a celebrity in the research world and many researchers know him by his name, Alex. Unfortunately, Alex died soon after these experiments and we will never know what the mechanism was that he used for addition. But if it is correct that the parrot 'counted up' in conducting the task – the scare quotes were wisely used already by Pepperberg – it would be an important finding. Instead of using only the proto-arithmetical capacities of OTS and ANS, the parrot would have been in some way manipulating numerical representations. Indeed, this kind of 'counting up' in doing addition is how children standardly first learn to add (Fuson, 1987). Alex could have been demonstrating an early form of abstract numerosity manipulation, in which case the term 'proto-addition' would be very much suitable.

Some cases of what McCrink (2015) calls 'intuitive non-symbolic arithmetic', like the Wynn doll experiment, seem unlikely to be even proto-addition in the sense that numerical representations are processed in any way as addends. Other cases, like Alex the grey parrot, can be seen more feasibly as proto-addition, even though more information would be needed to reliably assess the cognitive processes involved. Therefore, I do not want to dismiss the possibility of there being genuinely proto-arithmetical operations of addition, and perhaps also other operations. However, in general, I strongly disagree with passages like the following:

> From very early on in development, infants are sensitive to the effects of arithmetic operations that involve nonsymbolic number representations. Infants, children, adults, and nonhuman animals can productively combine and relate two magnitudes in an inexact manner. In humans, these skeletal abilities undergo substantial development across infancy and childhood, and on into adulthood, as children become enculturated and their abilities to represent both the magnitudes and their operational outcomes sharpen. (McCrink, 2015, p. 216)

From what we know, this is a flawed way of interpreting the data. It is not the case that we have an approximate ability to 'combine and relate magnitudes' that then gradually develops and sharpens in the process of enculturation. As I have argued in detail in Chapter 3, enculturation with natural numbers and their arithmetic is a transformative process, not one of gradual development where proto-arithmetical abilities 'sharpen' into arithmetical abilities. Therefore, when trying to reveal the cognitive development of basic arithmetical abilities, we should not look only into proto-arithmetic.

This is not to say that proto-arithmetical abilities are irrelevant for the development of arithmetic. Indeed, they appear to be relevant in at least two different ways. First, if the account presented in this book is along the right lines, proto-arithmetical abilities play a central role in number concept acquisition. Second, it is plausible that proto-arithmetical abilities also play some kind of cognitive role in grasping basic arithmetical operations. Grasping addition, for example, may be easier because the ANS already gives us a rough notion of how combinations of numerosities behave. It is also likely that, even as arithmetically educated individuals, we use the ANS in connection with arithmetical tasks. As seen in Section 1.3, we cannot turn off the proto-arithmetical abilities even when conducting symbolic arithmetical tasks. In that section, the examples of number comparison tasks showed that this can actually make our cognitive processes sub-optimal. But of course, in many cases this parallel activation of the ANS can be beneficial. If we make an error in our calculation and end up, for example, with the sum 66 + 85 = 111, it is possible that the ANS sends an error signal, because the sum is so far off from the correct one. In the case of 66 + 85 = 141, for example, we could only detect the error by arithmetical calculation.[2]

[2] It is also possible that it is partly simply experience with mathematical calculations that can make us detect errors. Most likely there are multiple factors in play. For more on this question of mathematical metacognition, see Wilson and Clarke (2004).

Nevertheless, while proto-arithmetical abilities influence the development of arithmetical abilities, our focus should be on the stages of development in which properly arithmetical abilities emerge, because they are more likely to mirror the cultural historical development of arithmetic. This means the stages at which arithmetical operations are grasped in a systematic, general and exact manner. For arithmetically educated individuals, mental as well as pen-and-paper arithmetic is conducted with the help of addition and multiplication tables that are usually simply memorised (see, e.g., Trivett, 1980). In a decimal system like that of the Indo-Arabic numeral symbols, this means memorising a hundred addition facts and another hundred multiplication facts.[3] In other numeral symbol systems, this is different. For example, in the Roman numeral symbol system, six simple simplification rules are enough for addition (Schlimm & Neth, 2008).[4]

Here it is not possible to focus on the learning of arithmetic generally across different cultures and different numeral systems, so I will focus only on the Indo-Arabic numeral symbols and standard 'Western' educational methods. In these methods, the addition and multiplication tables play a key role, but of course they are not the first introduction to arithmetical operations; the tables as such do not tell us what addition or multiplication are like as operations. For children to understand arithmetical operations, something else is needed. It is this 'something else' that we need to consider in relation to the phylogeny and cultural history of arithmetical operations. How do people originally come around to conducting arithmetical operations, after they have created number concepts? Views on this may vary and considerations of space do not allow for a comprehensive treatment of the topic. For the present purposes, however, such a wider treatment is not crucial. Indeed, just like with the development of number concepts, we are often firmly in the area of conjecture when it comes to understanding the phylogeny of arithmetical operations. What I hope to be able to provide here is one plausible account, supported by empirical data, of how arithmetical operations may have developed as a cultural practice. Here I will focus on addition, and only briefly discuss multiplication, subtraction and division.

[3] Although this number can be brought down by, for example, understanding the symmetricity of addition and multiplication.

[4] Schlimm and Neth write that seven rules are needed, but in fact six are enough: IIIII = V, VV = X, XXXXX = L, LL = C, CCCCC = D, DD = M.

In their account of the development of mathematics, Lakoff and Núñez (2000) see mathematics as applying a collection of *conceptual metaphors* where procedures conducted in our physical environment develop abstract, metaphorical counterparts. The metaphor they associate with arithmetic is that of 'arithmetic is object collection' (p. 55). The way a metaphor works according to conceptual metaphor theory is that there is a *source domain* and a *target domain*, as well as a mapping between them (Lakoff & Johnson, 2003). The mapping takes items from the source domain and maps them on the target domain. The general idea behind such metaphorical cognition is that we grasp the items and processes in the source domain, which then helps us grasp the mapped items and processes in the target domain. The standard example of this in the literature is the metaphor 'Love is war', in which *war* is part of the source domain, *love* is part of the target domain and the metaphor is apt in case our beliefs about war have enough structural similarity to our beliefs about love. In the case of arithmetic, in the account of Lakoff and Núñez (2000), the source domain is a physical object collection and the target domain is arithmetic.[5]

The processes related to arithmetic in the account of Lakoff and Núñez are very simple. Numbers, for example, are seen as the metaphorical counterpart of 'collections of objects of the same size' (Lakoff & Núñez, 2000, p. 55). Addition in the target domain of arithmetic is then seen as the mapping of 'putting collections together' in the source domain. Subtraction is 'taking a smaller collection from a larger collection' (Lakoff & Núñez, 2000, p. 55).[6] With multiplication things get a bit more complicated, but remain still relatively simple: the multiplication $A \times B = C$ is seen as the metaphorical counterpart of the process of 'the pooling of A subcollections of size B to form an overall collection of size C'. Division $C \div B = A$ is conversely 'the splitting up of a collection of size C into A subcollections of size B' (p. 61). This is one suggestion they make. The other one is that multiplication is 'The repeated addition (A times) of a collection of size B to yield a collection of size C' and division 'The repeated subtraction of collections of size B from an initial collection of size C until the initial collection is exhausted. A is the number of times the subtraction occurs' (p. 62).

[5] It should be noted that the conceptual metaphor theory of Lakoff and Johnson is far from being the only relevant theory of metaphor in the literature. For other approaches, see Kövecses and Benczes (2010) and Way (1991).

[6] Recall here that we are in the domain of natural numbers, so negative integers have not yet been introduced.

Is this what happened when people first started to use arithmetical operations? Did the first people with arithmetical abilities cognise metaphorically in this way, whether explicitly or implicitly? There is reason to be sceptical about the details. The latter explanation of multiplication, for example, sounds plausible, but it seems more unlikely that people would grasp the source domain process of the latter explanation of division. Indeed, it generally seems more likely that the way arithmetical operations are developed and learned is a more dynamic development of skills and knowledge, which cannot be captured neatly by the kind of metaphors suggested by Lakoff and Núñez. Moreover, I have previously argued that, if conceptual metaphors play a key role in the development of mathematical cognition, it is more likely that this concerns general metaphors instead of the kind of wide collection of specific metaphors that Lakoff and Núñez propose. The metaphor I have proposed in earlier work is the *Process → Object Metaphor* (POM), in which we treat the end products of processes, like that of counting, as an object, in this case a natural number (Pantsar, 2015b, 2018a).[7] This kind of general metaphor can then be connected to different source domain processes, also including the ones identified by Lakoff and Núñez. Object collection is clearly a process, and it is highly plausible that manipulating blocks or other items facilitates grasping the abstract arithmetical notion of addition. Certainly, it is not simply a coincidence that concrete objects are widely used in teaching addition in early education.[8]

But how, in fact, does the development of addition happen through a process involving physical objects? Here it is useful to twist the order of development around for a while and see how addition is defined in modern axiomatic approaches to arithmetic. In the Dedekind–Peano axioms (found in full on page 133), addition is defined recursively via the successor operator as follows (Peano, 1889): $a + 0 = a$, $a + S(b) = S(a + b)$. Here S is the successor ('the next number') operator and a and b are some natural numbers. Unfortunately there is no historical data concerning the origin of addition that could establish whether this kind of application of successor operator was involved. However, I believe that something can be said about that matter, given that it is plausible that the cultural development of addition is (at least to some degree) mirrored in the way children

[7] A similar metaphorical account of mathematical objects has been proposed by Sfard (2008).

[8] Indeed, this type of 'hands-on' educational practice of learning arithmetic through physical manipulation of objects has not always been around. The current practice emerged in the latter half of the nineteenth century and its standard work was the 'Quincy system' introduced by Parker (1879).

in that culture are taught arithmetic. Certainly, children do not learn addition through the kind of definition used in the Dedekind–Peano axioms, but how they learn it might actually be captured by it. What the definition says, essentially, is that you reach the desired sum $a + b$ by applying the successor operator b times to the number a. In other words, you *count* to b starting from a.

According to psychologists, this is indeed what children can do to learn addition. Fuson and Secada (1986) have distinguished between two ways of using counting to add when there are entities (e.g., blocks) present for each addend. In the 'counting-all' strategy of $a + b$, children count all the members of a (e.g., a group of blocks) and then continue the count when moving to the members of b. So the addition $8 + 5$ is counted as $1, 2, 3, 4, 5, 6, 7, 8, 9, 10, 11, 12, 13$ (Fuson & Secada, 1986, p. 130). A major developmental shift happens when children realise that they do not need to re-count the members in establishing the addition in case they have already counted them. So, if a child has counted that there are eight items in the collection a, they can do two things. They can either use the 'counting-all' strategy and recount them and continue the count to b. Or they can use the '*counting-on*' strategy and only count the items in b: $9, 10, 11, 12, 13$. The 'counting-on' strategy is conceptually more advanced and helps children grasp addition (Fuson & Secada, 1986, p. 130). Interestingly, it also mirrors the way addition is defined in axiomatic arithmetic. I do not believe that this is a coincidence. What the Dedekind–Peano axioms aim to do is to give as mathematically simple a definition of addition as possible. It is feasible that there is a connection between that definition and the *cognitively* simplest way of grasping addition of numbers. If this is the case, aside from individual ontogeny, it is plausible that also the development of addition on a cultural level has followed a similar path.

It is important to note that addition understood in this way is a direct descendant of counting. Thus, addition itself may not originally mean anything more than a particular type of counting procedure applied to two collections of items. This is in line with the Process → Object Metaphor. What the POM states is that we treat end products of processes as objects. This is what happens in counting: the end product of the process of reciting five numeral words in order is treated as an abstract object, the number five. Addition is a continuation of this. Ultimately, the addition $8 + 5 = 13$ is just first reciting eight numeral words in order and then continuing from that point with five more numeral words. The end product of the counting process, the word 'thirteen', is then treated as an object, that is, the result of the addition (the *sum*). When we learn addition tables, we no longer need to calculate the sum for single-digit numbers, and for multi-digit additions we use an algorithm applying the addition tables instead of counting. But originally, the way to learn addition can be seen as a direct continuation of counting. Multiplication can then be defined with the help of addition, and again the way it is learned as repeated addition mirrors the axiomatic definition.

Unsurprisingly, just like counting itself can be aided with embodied tools, such as fingers, embodied learning is advantageous for the learning of addition (Butterworth, 2005). It has been established, for example, that finger gnosis levels predict the level of calculating addition tasks with fingers in children of five-to-seven years old (Reeve & Humberstone, 2011). This is not to claim that using fingers is indispensable for learning to add. Other physical objects, like building blocks, can serve similar purposes (Verdine et al., 2014). But the evidence clearly points to the direction that addition is easier to grasp when connected to collections of physical items, thus mirroring the general framework set by Lakoff and Núñez.

The above considerations mainly concern ontogeny, but in this chapter we are interested in phylogeny and cultural history. I have suggested that it is plausible that these two aspects of the development of arithmetical abilities are connected. Thus, I believe that metaphorical accounts can help explain both the way individuals learn arithmetic and how arithmetical operations emerged originally in human cultures. But is there any historical record to support this view? For the most part, the record leaves us badly lacking in terms of concrete evidence. However, evidence from Egypt from around 4,000 years ago suggests at least a similar order of addition and multiplication, as presented in this section:

> The fundamental arithmetic operation in Egypt was addition, and our operations of multiplication and division were performed in Ahmes's day through successive doubling, or 'duplation'. Our own word 'multiplication', or manifold, is, in fact, suggestive of the Egyptian process. A multiplication of, say, 69 by 19 would be performed by adding 69 to itself to obtain 138, then adding this to itself to reach 276, applying duplation again to get 552, and once more to obtain 1104, which is, of course, 16 times 69. Inasmuch as $19 = 16 + 2 + 1$, the result of multiplying 69 by 19 is $1104 + 138 + 69$, that is, 1311. (Merzbach & Boyer, 2011, p. 12)

This passage concerns how multiplication emerged from addition, but how did addition emerge in Egypt? Unfortunately we cannot know, but

I propose that the ontogenetic account I have outlined in this section is at least a plausible phylogenetic explanation for any culture in developing arithmetical operations. The idea is that the individual ontogeny of arithmetical abilities at least partly recapitulates the phylogeny and cultural history of arithmetic. This is a plausible view because, as we have seen, in addition to cultural factors, the ontogeny of arithmetic is constrained by proto-arithmetical abilities. The proto-arithmetical abilities are similar across populations, so it is plausible that they have been applied in similar ways in acquiring number concepts throughout the cultural history of arithmetic. The cultural factors, on the other hand, determine in an important way the manner in which children learn about number concepts and arithmetic. In this way, the way arithmetic is taught is likely, at least to some degree, to mirror the way arithmetic has developed.

For more complex arithmetical operations this connection between ontogenetic acquisition and cultural development may be looser, given that the historical trajectory is longer. But this is different for addition. Once a culture processes number concepts and can use them for counting, there is only a small step from that to addition. Addition, as we have seen, is at its simplest continuing counting from one collection to another. We have no surviving record of how cultures have made this step, but perhaps the absence of evidence in this case suggests something informative: addition may have been such a straight-forward continuation of counting that its origins were not thought to require any written explanation. What we do know is that mathematical texts from cultures that developed arithmetic independently, from China (Needham & Wang, 1995) to the Mayans (Kallen, 2001), show how to solve addition problems. However this happened, we know that the development of numeral symbols led to the development of addition and consequently also multiplication.

This is not to suggest that addition can only be conducted when there is a system of numeral symbols in place. The Oksapmin people of Papua New Guinea, for example, have an extensive body part system for quantities. They have shown to be fast to learn to solve addition problems when interviewed by researchers (Saxe, 1982). This may suggest that, even though they do not have an extensive system of numeral words or symbols, they may possess some understanding of proto-addition. As discussed earlier in this section, the existence of non-symbolic proto-addition abilities is perfectly compatible with the approach here. I am not claiming that approximate sums, for example, cannot be established without possessing exact number concepts. However, for arithmetical operations as we understand them, we need to process number concepts. It also seems that, once we do process number concepts, addition with them comes relatively easily, followed by multiplication and other arithmetical operations. Therefore, the development of basic arithmetic in cultures can be seen as a continuous development from number concept acquisition. Indeed, it is likely that number concepts themselves reach an important status in a culture *because* they allow conducting arithmetical operations which, in turn, can provide useful in a variety of different applications.

5.2 Arithmetical Applications

In Section 5.1, we have seen how basic arithmetic, starting from addition, can be a continuation of number concept acquisition in phylogeny and cultural history. Without cognitive tools, however, what we can do with arithmetic is very limited. That is why the development of arithmetic is closely connected to the development of symbol systems, writing tools, abaci and many other tools to assist the cognitive processes involved in arithmetic. In this regard, the place-value principle of the Indo-Arabic numeral system, for example, is highly important. The place-value principle states that the numerical value of a symbol depends on its place in a complex numeral. Thus, the numeral symbol 9, for example, refers to a different value in numeral 192 than it does in 912. This principle made possible the standard Western schoolbook algorithms for multiplication and division, and it also enables determining the magnitude of a number by a glance at the numeral. Due to the place-value principle, we can also immediately know that 3012 is larger than 832 because the former numeral has more digits. This information is not present, for example, in the Roman numerals MMMXII and DCCCXXXII.

But, perhaps more importantly, the numeral symbol system also influences how easy it is to calculate arithmetical operations (Detlefsen et al., 1976). Since the process of completing the same task varies as a function of the employed numeral symbol system, an expected consequence is that a different profile of cognitive capacities (e.g., working memory, semantic fact retrieval and visual attention shifts) is applied (Schlimm & Neth, 2008). With Roman symbols, easy additions like XXV + XVII = XXXXII are almost immediately solved by a straight-forward grouping of the symbols (and applying one simplification rule), while with Indo-Arabic symbols, the task 25 + 17 = 42 requires memorising several addition facts, as well as applying the place-value principle (and carrying). However, as the numbers become larger, the amount of manipulations required by applying the Roman symbol system increases. Let us take, for example, the multiplication 235 × 28 = 6580 conducted with Roman symbols. By using a multiplication table for each partial product (e.g., C × XXVIII = MMDCCC) and adding them together, we get the following calculation:

CCXXXV × XXVIII

= MMDCCC MMDCCC CCLXXX CCLXXX CCLXXX LLXXVVVV

= MMMMDDCCCCCCCCCCCLLLLXXXXXXXXXXXVVVV

= MMMMMMCCCCCLXXX

= MMMMMMDLXXX

It is easy to see how the completion of multiplication tasks with Roman symbols becomes more time-consuming and error-prone as the numbers become larger. Indeed, it would be tempting to deem Roman symbols as inferior cognitive tools for multiplication tasks (as is done by, e.g., Zhang & Norman, 1994). However, for an individual enculturated with Roman symbols, this kind of calculation may not have seemed particularly difficult. Indeed, it is possible that, for an individual proficient with Roman numeral symbols, memorising all the multiplication and addition facts necessary for using the standard Indo-Arabic schoolbook multiplication algorithm would have seemed like a considerably more difficult prospect.[9]

It is therefore important to realise that the development of arithmetical cognition is always tied to the particular culture and the practices and tools that are used. This is the case when we consider calculating results of arithmetical operations, but it is also important when we consider the role of arithmetic in particular cultures. This has ranged from cultures with no arithmetic (such as the Pirahã and the Munduruku) to cultures where arithmetic was held in high regard (such as the Mayans and Ancient Chinese) to large parts of the modern world where almost all children are taught arithmetical skills. In the modern industrialised and monetised world, it is easy to understand why arithmetical skills are important. In the Western world, for example, we need to be able to handle financial affairs, keep track of time, understand measurements of distances, weights, and so on; all skills without which living in the modern world becomes highly difficult, if not impossible.

[9] For more on the role of numeral symbols in learning and practicing arithmetic, see Schlimm (2018) and Stjernfelt and Pantsar (2023). For a wider comparative analysis of different numeral symbol systems, see Chrisomalis (2010).

All these modern needs, however, are *results* of humans having arithmetical skills. In the Western world, we need to possess arithmetical skills and knowledge because we have structured our societies in thoroughly arithmetical ways. But why did this development take place? Why did arithmetic become such an important backbone of our societies that it is considered fundamental for basic education, as shown by the phrase 'three Rs' used widely in the United States (reading, writing and arithmetic – or 'rithmetic, rather). This is the question I will focus on in this section, and I believe the key to any answer is found in the way arithmetical *applications* emerge.

The first known application of number symbols can perhaps be found in bookkeeping, given that, according to the theories presented in Section 4.2, this is how number symbols were originally born. It is easy to see how bookkeeping could also quickly make use of arithmetical operations conducted on those symbols. Bookkeeping with numeral symbols facilitates trade and the storage of goods. Exact inventories can be kept and volumes of trade can be recorded and communicated. This most likely also had an effect on agriculture, which also benefitted from the invention of arithmetic in other ways. Land areas and crops could be calculated and numerals allowed for more exact timekeeping in the form of calendars. The list could go on: from navigating to religious practices, the introduction of numerals transformed the way many societies functioned (Everett, 2017).

However, it is important to note that there were significant differences in the applications that arithmetic found in different cultures. Indeed, perhaps it is the lack of important applications that partly explains why some cultures did not develop arithmetic. The Pirahã, for example, are a hunter-gathering tribe that engages in trade only in a limited way. While arithmetic may have proven to be useful for them, it could also be that there was no immediate need for the advantages brought on by numeral systems. Indeed, given the possibility that the ancestors of the Pirahã had a more extensive numeral system, it could be that the lack of need for numerals made them slowly forgotten in the Pirahã culture.[10]

[10] This possibility gets some support from the fact that many cultures that diverged from the Pirahã during the peopling of South America have numeral words in their languages (see, e.g., Ifrah, 1998). It could be that they developed numeral words after diverging from the Pirahã but, given the relative recency of this divergence, it is possible that the culture from which the Pirahã diverged had a language with numeral words. To the best of current knowledge, the first humans reached Brazil around 10,000 BC (Posth et al., 2018). It is not known when exactly the Mura people, of which the Pirahã are the only surviving subgroup, diverged as their own lineage.

In other cultures, however, numeral words, numeral symbols and arithmetic clearly found use. Most likely this happened in a variety of ways, which can be seen in the different trajectories that the development of arithmetic and arithmetical applications took. The surviving arithmetical texts from Ancient Egypt, for example, show many problems related to sharing products like bread and beer (Merzbach & Boyer, 2011, p. 13). While there must have also been many other applications of arithmetic, it seemed to play an important role in everyday life. It was also used in geometrical applications, although this was done to a greater extent during roughly the same period in Mesopotamia. It is important to note, however, that, to the best of our knowledge, neither in Egypt nor in Mesopotamia was there a clear and explicit distinction between approximate and exact results in place. In this sense, they differed importantly from Greek mathematics, in which it was crucial to find exact results to geometrical problems, like those of calculating areas or volumes. While pre-Hellenic mathematics was certainly not only a practical discipline, there seemed to have been a clear focus on practical applications and less on explicit rules and distinctions (Merzbach & Boyer, 2011, 38–39).

The surviving mathematical texts from Egypt and Mesopotamia consist mainly of specific problems and their solutions, and the same is true of Ancient Chinese mathematics, which dates at least to 300 BC. The Chinese numeral symbol systems are even older and may have been present already in 1,300 BC (Ifrah, 1998, p. 269). Interestingly, however, the numeral symbols were not, to the best of our knowledge, used for arithmetical calculations. They were used for expressing quantities, but for calculations an early version of the abacus, with little ivory or bamboo rods and a tiled surface like a checkerboard, was used, setting a long tradition of calculating devices in China that culminated in the 'Chinese abacus' we are currently familiar with (Ifrah, 1998, pp. 283–288). These abaci enabled incredibly fast arithmetical operations, which made them highly important in commerce – an influence that is still present in China and Japan. However, while this focus on calculations and practical problems may give the idea that Ancient Chinese mathematics was mostly applied, this would be a hasty conclusion. The Ancient books, like the highly influential *Nine Chapters on the Mathematical Art*, are focused on problems related to practical applications, including agriculture, engineering and taxation. But Ancient Chinese mathematicians seemed also to have solved problems for the sake of solving problems. Mathematics, including arithmetic, in China also developed a more theoretical side which was not directly connected to particular applications (Merzbach & Boyer, 2011, p. 181).

The final strand of arithmetical development I want to discuss in this section is the Mayans. During the first millennium CE, the Mayan civilisation had a system of numeral symbols and arithmetic which they could use for astronomic measurements with astounding precision. They calculated the synodic revolution of Venus, for example, at 584 days, when the modern calculation is 583.92. Their calculation of the solar year on Earth was 365.242, which is exactly what the modern computations give (up to the third decimal) (Ifrah, 1998, p. 297). However, as far as we know – and tragically we know very little since, along with remains of the Mayan civilisation, most of their writings were destroyed by the Spanish invaders – arithmetic was not widely used for everyday applications. Numeral symbols were used primarily not by accountants, but by priests as arithmetical calculations and astronomy were tightly connected to mysticism and religion. Calendar keeping in particular was highly important for the astronomer priests and it led, for example, to an early introduction of the number zero and a place-value system of numerals (Ifrah, 1998, p. 311). For calculations of arithmetical operations, the Mayan arithmetic was highly developed. It seems to have had very different practical applications from what was happening elsewhere in the world, yet the arithmetical content itself developed largely on a converging path, in the sense that the same arithmetical operations were used to get the same results.

Indeed, arithmetic developed on largely converging paths to a large degree everywhere where it was developed. As far as we know, the Mayans, for example, did not have the notion of infinity. They did have the number concept zero, however, which was not present in Ancient Greek mathematics. Yet the Greeks did develop the notion of infinity (more about this in Section 5.3). What we can see is that arithmetic took on particular trajectories in different cultures, in a development that was most likely shaped by the applications there were for arithmetic. But, regardless of the applications, when it comes to basic operations (addition, multiplication, subtraction, division) of non-zero natural numbers, arithmetic converged even when developed independently.

In Part III of this book, this observation will be put to important use. Clearly arithmetic cannot be simply a matter of conventions if cultures develop it independently of each other and end up with highly similar results. There is something more to arithmetic, something that makes arithmetic more *objective*. But, before that, we need to know more about how arithmetic has developed into its modern form. So far, we have been considering basic arithmetic and applications of arithmetical operations. That, however, is not what modern mathematicians would associate with arithmetic. For them, arithmetic is essentially a formal system of proving theorems. Next, we will see how this development was possible.

5.3 Proofs, Formalisation and Infinity

As we have seen, even though arithmetic has independently developed in converging ways in different cultures, there have also been important differences. Aside from particular differences, like the number symbol for zero, one main difference concerns the *aim* of arithmetic. We have seen that, in Egypt, Mesopotamia, China and the Mayan civilisation, the aim of arithmetic was mostly to solve different types of problems by calculation. However, this is only a small part of what we would currently consider to be the purpose of arithmetic. In the present 'Western' mathematical tradition, arithmetic as an academic discipline is above all about proving general theorems concerning all natural numbers, or all numbers of a particular infinite subset (like those of even or odd numbers).

The standard story in the history of mathematics was for a long time that proving general theorems became part of mathematics in Ancient Greece, as evidenced above all by Euclid's *Elements* and the writings of Archimedes.[11] However, as shown by Karine Chemla, this history is highly misleading. New analyses of Indian, Chinese, Egyptian and Arabic source materials reveals that proofs had been part of mathematics much more widely, and in a much important way than previously thought (Chemla, 2015. p. 7). As argued by François Charette, when writing histories of mathematics in the nineteenth and early twentieth centuries, there was a tendency to dismiss any 'Oriental' mathematics as 'imaginative', whereas Greek mathematics was emphasised as 'logical' (Charette, 2012). No doubt this was partly due to damaging cultural imperialism in which mathematics was promoted as a 'Western' invention and contrary evidence was often simply dismissed. This, of course, is made doubly absurd by the fact that the nineteenth century Western world cannot be in any feasible way seen historically as a cultural continuation of Ancient Greece. In reality, much of what we value as the great intellectual achievements of Ancient Europe had to be re-introduced to Europe via the Orient (see, e.g., Frankopan, 2016).

It is therefore important to note that mathematical proof as a notion should not be historically associated exclusively with Ancient Greece.

[11] The earliest known evidence of Ancient Greek mathematics dates to the second half of the fifth century BC, to Hippocrates' quadrature of lunules (Netz, 2004).

Nevertheless, in this section I will focus on the Greek tradition. The reason for this is simple: it is the surviving material from Ancient Greece that gives us the best access to how arithmetic developed into a formalised discipline in which general proofs were conducted. It may well be that the Greek mathematicians had more focus on clear distinctions and exact solutions than was the case elsewhere, as suggested by Boyer (1991). But it could also be the case that the surviving writings are not representative of the variety of mathematical activity that was taking place in the Ancient world. What is beyond doubt, however, is that in Ancient Greece the notion of mathematical proof emerged in a way that influenced modern mathematics enormously.

Mathematical proof in Ancient Greece is mostly associated with geometry and in particular the book *Elements* by Euclid (Euclid, 1956). The notion of Euclidean proof, however, is quite different from the modern formal notion of 'proof' in mathematics. Instead of conducting a line-by-line proof of logical steps, Euclidean geometry is heavily based on a particular method of proof in which lettered diagrams play an indispensable role (Manders, 2008; Netz, 1999). Because of such important differences in the methodology, Ferreirós (2016) has argued, we should not conflate the notion of axioms and proof in Euclid with modern notions, like those used in Hilbertian geometry (Hilbert, 1902). Rather, Ferreirós argues that Euclid's postulates should be thought of as rules for constructing diagrams.[12]

While it is important to point out the difference between the modern notion of 'mathematical proof' and that present in Euclid's work, this difference should not be over-emphasised. The most important ingredients of mathematical proof were already in place in Euclid. Starting from generally accepted axioms, we can derive theorems through logical steps of proof. When this process is conducted properly, the truth of the theorems is undisputed. Mathematical practices have of course evolved, but ultimately this notion of proof has remained as the guiding principle in modern mathematics as well.

Most of the thirteen books of Euclid's *Elements* concern geometry, but books seven to ten are about arithmetic. In them, Euclid collects the arithmetical knowledge of the time (ca. 300 BCE), including the famous algorithms for finding the greatest common divisor and the least common

[12] While the point about different notions of proof remains valid, it should be noted that the status of diagrammatic proofs is debated currently among philosophers, many of whom accept the idea that a diagram can provide a valid proof (see, e.g., de Toffoli, 2023).

multiple. The approach is remarkably similar to the one found in modern mathematics textbooks, starting from basic notions related to numbers (e.g., primeness and divisibility) and then moving on to more difficult topics. In Proposition 20 of Book IX one can find perhaps the most important arithmetical proof of the Hellenic times, one which is still widely taught to mathematics students. This proof of the infinitude of prime numbers came to be known as Euclid's theorem.

The proof is beautiful in its simplicity. Let us consider any finite list of all prime numbers $p_0, p_1, p_2, p_3, \ldots, p_n$ and let P be the product of all the numbers on the list, that is, $P = p_0 \times p_1 \times p_2 \times p_3 \times \ldots \times p_n$. Now let us consider the number $q = P + 1$. Clearly q must be either a prime or not. If it is a prime, then there must be a prime that is not on the list, namely q. If q is not prime, then it must be divided by some prime number p_i. If p_i is on our list, then it divides P. However, by our assumption, p_i also divides the number $q = P + 1$. Thus p_i must also divide 1, which is impossible for any prime number. Therefore, we establish that q cannot be on the list of prime numbers. Given that the list was arbitrary, we have proved that no finite list can contain all prime numbers, that is, there are infinitely many primes.

The proof is valid and it, or one of its reformulations, is probably familiar to anyone who has studied mathematics at a university level. However, it is important to note that the proof presented here is heavily paraphrased in relation to Euclid's work. Euclid did not have a method of writing arbitrary lists of primes. Instead, as elsewhere in *Elements*, he picked sample elements, three primes with which he showed that another prime can always be found. It is not specified that, instead of three primes, we could pick four, five or any other number of primes. Therefore, while Euclid's theorem is valid and proofs of it follow Euclid's proof of his Proposition 20, it is important to note that Euclid's original proof was not strictly speaking rigorous in the modern sense. Mathematical proof did not appear out of nothing into Ancient Greece and neither was it developed there to its modern form.

Nevertheless, Euclid's *Elements* was still a remarkably modern book, and the proof of the infinity of primes was one of the great achievements of Ancient arithmetic. Indo-Arabic numeral symbols facilitated arithmetical calculations, but it took a long time before arithmetic was developed substantially beyond what was present already in Euclid. The culmination of this later development happened in the nineteenth century through the work of Dedekind (1888) and Peano (1889). In the latter, Peano used

Dedekind's work to present arithmetic as formal axioms, which have since become known as Dedekind–Peano axioms or simply Peano axioms. The Peano axioms have the following content:

1. 0 is a natural number.
2. For every natural number n, $n = n$.
3. For all natural numbers n, m, if $n = m$, then $m = n$.
4. For all natural numbers n, m, p, if $n = m$ and $m = p$, then $n = p$.
5. For all x, y, if y is a natural number and $x = y$, then x is a natural number.
6. If n is a number, the successor of n, $S(n)$ is a number.
7. For all natural numbers n, m, $m = n$ if and only if $S(m) = S(n)$.
8. For every natural number n, $S(n) = 0$ is false.
9. Let φ be a unary predicate. If $\varphi(0)$ is true and if $\varphi(n)$ being true implies that $\varphi(n+1)$ is true, then $\varphi(n)$ is true for every natural number.

The first axiom determines what the first number is. Sometimes this is also given as one, and Peano himself used both one and zero on different occasions. The axioms 2–5 define the equality relation of natural numbers.[13] The axioms 6–8 define the arithmetical properties of natural numbers and finally, the axiom 9, the so-called axiom of *induction*, establishes how we can prove things concerning all numbers. The axiom of induction can be presented as a second-order axiom (as in here) or a first-order axiom scheme. In mathematical logic, this distinction is highly important, but for the present purposes we do not need to be concerned with that (for details, see Pantsar, 2009).

The Peano axioms are clearly a long way from the early cultural origins of arithmetic. But a close look at the content of the axioms tells us that they formalise ideas that had been present for millennia. The first axiom only states that there is at least one natural number. The axioms concerning the equality relation formalise the simple idea of two numbers being equal, another idea that had been around for millennia. The arithmetical properties of numbers set by axioms 6–8 have also been around in counting lists for a long time. Axiom 6 is equivalent with every numeral word having a unique numeral word next on the list. Axiom 7 states that we cannot arrive at different numeral words by taking the same step in the

[13] Nowadays these axioms are often unmentioned, since they already follow from first-order logic with equality, which (or a stronger logic) is used as the logical language of arithmetic (for more, see Pantsar, 2009).

counting process. Axiom 8 is then equivalent with there being a beginning to the counting list of numeral words, which is a universal characteristic of counting lists. Therefore, the first eight axioms formalise ideas that have been behind pre-formal arithmetic wherever it has been developed.

It is only the axiom of induction that does not seem to be universal to arithmetic. It was implicitly in place in Euclid's methodology, with the sample numbers effectively playing the role of arbitrary lists of numbers. But, as far as we know, such an idea for general proofs about properties of numbers was not – at least explicitly – present in all arithmetical cultures. From the surviving evidence, it appears that, for all their great advances in arithmetic, the Mayans, for example, did not prove theorems concerning all natural numbers. From the point of view of the Greek tradition, this may seem odd. Why was there such a difference? As we have seen, Mayan arithmetic seems to have been closely connected to astronomy and religion. Perhaps this was even their main purpose of arithmetic. In Ancient Greece, the development took a different path. Possibly at least partly due to the influence of Plato, aside from practical applications, mathematics was seen as a pursuit of eternal truths that were valuable in their own right (Heath, 1921).

This might also explain one important difference between Greek and Mayan mathematics. While the Mayans, as far as we know, did not develop an explicit notion of infinity, for Ancient Greek mathematics it was central. In explaining the cognitive origins of arithmetic, infinity is a particularly interesting topic. If arithmetical cognition is rooted in physical interactions within our environment, we must ask how these clearly finite interactions in a finite environment can lead to the idea of infinity of numbers.

How did infinity enter mathematics? Again, the origins are lost in history, but some conjectures have been presented. Lakoff and Núñez (2000), for example, evoke a special *basic metaphor of infinity* to account for the introduction of infinity into mathematical cognition. As they point out, infinity in mathematics is not simply the negation of finiteness:

> [The negation of finiteness] does not give us any of the richness of our conceptions of infinity. Most important, it does not characterize infinite things: infinite sets, infinite unions, points at infinity, transfinite numbers. To do this, we need not just a negative notion ('not finite') but a positive notion – a notion of infinity as an entity-in-itself. (Lakoff & Núñez, 2000, p. 155)

The idea behind the basic metaphor of infinity is that processes that continue indefinitely are thought to have an ultimate result. Just like we speak of the final resultant state of a completed iterative process (such as counting to five), with the help of the basic metaphor of infinity we can speak of the 'final resultant state' of iterative processes that do not end. For example, the definition that the successor of each natural number is a natural number (the Peano axiom 6) gives rise to an indefinite iterative process. Once we realise that the process does not stop, we no longer expect a final resultant state. Rather, according to Lakoff and Núñez, we evoke a metaphorical 'final resultant state', which is the concept of actual infinity we use in mathematics. Just like the end products of completed iterative processes, this 'final resultant state' is unique and follows every non-final state of the indefinite iterative process (Lakoff & Núñez, 2000, pp. 158–159).

This way, Lakoff and Núñez try to explain that there does not need to be any further cognitive step involved in grasping infinity. We only need to grasp that some processes continue indefinitely and apply the basic metaphor of infinity. Could this ontogenetic account also explain how the notion of infinity was originally invented in, or introduced to, mathematics? I have serious doubts about that, but also about the ontogenetic account of Lakoff and Núñez itself. To me it seems unlikely that our basic cognitive toolbox includes anything specific to infinity. The case of the Mayans, for example, suggests that not even all arithmetical cultures develop a notion of infinity, or at least not in connection to arithmetic.

Yet I think that the idea that infinity is the result of metaphorical cognition is appealing. But instead of an infinity-specific metaphor, I have proposed that the same phenomenon can be explained by evoking the general Process → Object metaphor. What the POM enables is the treatment of end products of processes as abstract objects. Equipped with the understanding that some processes do not have end points, we can treat their metaphorical end products as infinite objects. This, however, is not some basic cognitive capacity that we have. Instead, it is a culturally developed idea that took a great deal of mathematical development to make explicit. Not surprisingly, it takes children a relatively long time to understand the mathematical notion of infinity, even after they have become proficient in arithmetic (for details, see Pantsar, 2015b).

Applying the POM to knowledge about unending processes, it is possible to grasp a mathematical notion of infinity, and this is captured by the Dedekind–Peano axioms. Peano's axiom 6 clearly evokes such a process without an end. For each number, we can always find the next number, which is the foundation for grasping that the set of natural numbers is infinite. But, of course, this knowledge was already present in Euclid and most likely much before. Indeed, that is how recursive numeral word and symbol systems work: by understanding the numeral word or symbol system, we can immediately grasp that (in principle) there is no limit to counting to larger and larger numbers.

However, from this we only get the notion of *potential* infinity, what Aristotle called *apeiron dunamei* (απειρον δυναμει). What mathematics needs is a notion of *actual* infinity, *apeiron hos aphorismenon* (απειρον ηοσ απηοpισμενον) (*Physics*, Book 3, chapter 6). This actual infinity is what Cantor (1883) brought to mathematics, while also showing that some infinite sets are larger than others. This goes beyond arithmetic and it is not possible to discuss different transfinite numbers here.[14] However, we have seen that starting from proto-arithmetical abilities, it is possible for cultures to develop arithmetic to include a notion of actual infinity. This does not demand any cognitive abilities specific to infinity. Instead, I have argued that the Process → Object metaphor applied to indefinitely continuing processes is sufficient.

This development of infinity as a mathematical notion has not taken place universally, which may be connected to different metaphysical ideas concerning the world.[15] However, in the development traced in this chapter, infinity is a continuation of an idea (recursivity) that has been present in numeral word and symbol systems everywhere where arithmetic has been known to be developed. Thus, we have seen that starting from proto-arithmetical abilities, by co-evolving numeral words and number concepts, cultures can develop arithmetical operations, the notion of proof and the concept of actual mathematical infinity. What needs to be explained next is *how* such developments are possible, that is, how are cultures able to develop knowledge and skills through millennia in the way it has happened in the case of arithmetic?

5.4 Summary

While in Chapter 4 the focus was on the cultural development of number concepts, in this chapter I moved to the development of arithmetical operations. I reviewed empirical literature and argued that addition has

[14] 'Transfinite number' was a notion introduced by Cantor to be able to treat numbers that are not finite but neither are they *absolutely* infinite, in the sense that they are greater than any number (whether finite or transfinite).

[15] For example, it is possible that the Mayan metaphysics were based on the notion that everything is finite or cyclical, as was famously the case with their idea of time (Tedlock, 1992). In the Christian tradition in which Cantor worked, infinity was often connected to the idea of an infinite God. This was a connection that Cantor (1883) himself took very seriously.

plausibly developed as a direct continuation of counting procedures, which in turn made it possible to develop other arithmetical operations (multiplication, subtraction and division). After that, reviewing literature on different cultures, I showed the importance of arithmetical applications for the development of arithmetic – or the lack of it. While arithmetical operations have developed in similar ways in different cultures (when arithmetic was indeed developed), in the final section I showed how some aspects of Western arithmetic (e.g., proofs, axiomatisations and infinity) are more culturally specific.

CHAPTER 6

Cumulative Cultural Evolution

In Chapter 5, I have considered the question of how arithmetical skills and knowledge can develop in a cultural context. However, what has not yet been discussed is how such cultural transmission of skills and knowledge is possible in the first place. In Part I of this book, I promoted the framework of enculturation as an explanation of how arithmetical knowledge and skills can be acquired in ontogeny. In this chapter, I now ask the same question concerning cultural history. How can we generally explain the cultural development of skills and knowledge like those involved in arithmetic? Given that the cultural history of arithmetic shows that humans have developed new cognitive abilities, an important part of this question is *how* such development can take place. How can new innovations like arithmetic emerge and become established within cultures? In this chapter, I will focus on these questions.

6.1 What Is Cumulative Cultural Evolution?

While particular innovations in the history of arithmetic are doubtlessly due to individual persons or traceable back to a small number of individuals, as is at least partly the case with the Dedekind–Peano axioms, generally we cannot expect to find such individual sources for cultural innovations.[1] My main starting point here is that cultural developments frequently take place in small (trans-)generational increments which we mostly cannot trace to particular individuals. This is the principal idea behind the notion of *cumulative cultural evolution* as proposed in different forms by many researchers (Boyd & Richerson, 1985, 2005; Henrich,

[1] This is not to suggest that the work of Dedekind and Peano was not influenced by previous mathematicians. Indeed, the entire idea of axiomatisation of a field of mathematics can be traced back to Euclid, and beyond. In addition, their work on arithmetic built upon other contemporary mathematicians. For more, see Merzbach and Boyer (2011).

2015; Heyes, 2018; Tomasello, 1999; for an overview, see Mesoudi & Thornton, 2018).

In processes of cumulative cultural evolution, culturally developed traits are transmitted across generations. In the literature, the focus is often on the cultural development of artefacts, but such culturally developed traits can also include skills, practices and knowledge, among other cultural innovations. Many researchers argue that, even though animals engage in social learning and even cultural traditions, this kind of cumulative cultural evolution is a uniquely, or at least almost uniquely, human phenomenon (Boyd & Richerson, 1996; Tomasello, 1999). Robert Boyd and Peter Richerson, for example, argue that culture, defined as variation acquired and maintained by social learning, is common among non-human animals. But this only rarely turns into the cumulative evolution of cultural traits (Boyd & Richerson, 1996).

Michael Tomasello argues that this is due to the 'cultural ratchet' effect: cumulative cultural evolution works like a ratchet, where the movement of the ratchet prevents it from reverting to an earlier position. According to this metaphor, human cumulative cultural evolution differs from animal cultures and social learning in that human cultures do not revert back to earlier, less developed states, as has been noted to be the case in, for example, primates (Kummer et al., 1985; Tomasello, 1999). Tomasello describes the essence of cumulative cultural evolution as follows:

> Basically none of the most complex human artifacts or social practices – including tool industries, symbolic communication, and social institutions – were invented once and for all at a single moment by any one individual or group of individuals. Rather, what happened was that some individual or group of individuals first invented a primitive version of the artifact or practice, and then some later user or users made a modification, an 'improvement', that others then adopted perhaps without change for many generations, at which point some other individual or group of individuals made another modification, which was then learned and used by others, and so on over historical time in what has sometimes been dubbed 'the ratchet effect'. (Tomasello, 1999, p. 5)

This kind of cumulative cultural evolution has been important because it has allowed humans to fully apply the force of cultures as enduring trans-generational units where traits are efficiently transmitted. In non-human cultures, innovation (e.g., the introduction of new practices) happens, but it does not endure and cumulate in a similar way over generations:

> In contrast, human cultures do accumulate changes over many generations, resulting in culturally transmitted behaviours that no single human

> individual could invent on their own. . . . What is amazing is that the same brain that allows the !Kung to survive in the Kalahari, also permits the Inuit to acquire the very different knowledge, tools, and skills necessary to live on the tundra and ice north of the Arctic circle, and the Ache the knowledge, tools, and skills necessary to live in the tropical forests of Paraguay. There is no other animal that occupies a comparable range of habitats or utilizes a comparable range of substinence techniques and social structures. (Boyd & Richerson, 1996, p. 80)

Here I want to pursue the idea that the phenomenon of cumulative cultural evolution can help explain the cultural history of number concepts and arithmetic. Arithmetic involves knowledge, tools and skills that enable human cultures to survive and thrive in different ways. The Pirahã and the Munduruku, for example, have been able to endure as cultures without number concepts or arithmetic. But their cultures have different characteristics from the ones that did develop arithmetic. As we have seen, agriculture, for example, was one key practice that was connected with the development of arithmetic. As hunter-gatherer cultures, neither the Pirahã nor the Munduruku engage in agriculture. Similarly, trade and the related bookkeeping were identified as important early applications of arithmetic. The Pirahã and the Munduruku, however, engage in trade only to a limited extent (Everett, 2017).

Such considerations can make us understand what kind of developments the emergence of number concepts and arithmetic are connected to. But they can also make us understand how cumulative cultural evolution takes place in arithmetical cultures. The notion of 'cultural ratchet' of Tomasello cannot mean that human cultures *never* revert to earlier states. Indeed, it is possible that this is exactly what happened with the Pirahã and the Munduruku. Given that those cultures are relatively recently detached from other South American cultures that do possess numeral words, it is feasible that they once possessed numeral words but later lost them. If this is true, it is most likely the result of a lifestyle where numeral words stopped being useful. Just like necessity can be the mother of invention, lack of need can result in cultures losing certain traits.

Why did many cultures continue to develop their numerical skills and knowledge, unlike the Pirahã? Based on the considerations of the previous chapter, it appears obvious that this was because they were in different ways *useful* for the particular cultures. Everett (2017, p. 116) recounts the words of the historian John Hemming who remarked that the non-arithmetical Munduruku were 'easily duped' by traders who sold their goods to them up to a four-fold mark-up. The traditionally mostly isolationist Munduruku had probably not run into this problem regularly. But for a culture that regularly engages in trade and is dependent on it for survival, being duped in trade would be a threat to survival. Thus, arithmetical skills and knowledge, just like the abilities to hunt and get shelter in other conditions, may have become existential matters for cultures as practices like agriculture and trade developed.

6.2 Cultural Learning

In Section 6.1, I have argued that the theoretical framework based on the phenomenon of cumulative cultural evolution can be useful for explaining the development of arithmetic. Understanding cumulative cultural evolution requires, however, understanding how the learning of new skills and knowledge takes place in a cultural setting. This process of individuals within a culture learning and passing on information is often called *cultural learning* (e.g., Heyes, 2012, 2018; Tomasello et al., 1993). But what exactly are the cognitive processes that comprise cultural learning? Are they genetic or cultural adaptations, or perhaps a combination of both? How is cultural learning possible and how can it be improved? And, most importantly for the present context, can cultural learning explain how numerical and arithmetical cognition have developed through a process of cumulative cultural evolution?

In their influential account, Tomasello et al. (1993) identified three ways in which cultural learning is manifest in human ontogeny. These are *imitative* learning, *instructed* learning and *collaborative* learning. Abilities in these forms of cultural learning, they argued, are correlated with social cognition, that is, cognitive processes involved in the social interactions of (standardly) conspecific individuals. However, among forms of social learning, in their account cultural learning is special in that it is uniquely human and allows for a greater scope and fidelity of transmission of behaviours, skills and information (Tomasello et al., 1993).

This distinction of cultural learning as a special type of social learning is also made by Joseph Henrich, who has described the emergence of social learning as follows:

> ... *social learning* refers to any time an individual's learning is influenced by others, and it includes many different kinds of psychological processes. *Individual learning* refers to situations in which individuals learn by observing or interacting directly with their environment, and can range from calculating the best time to hunt by observing when certain prey emerge, to engaging in trial-and-error learning with different digging tools. So,

> individual learning too captures many different psychological processes. Thus, the least sophisticated forms of social learning occur simply as a by-product of being around others, and engaging in individual learning. (Henrich, 2015, pp. 12–13, original emphasis)

This kind of social learning should then be distinguished from cultural learning, which requires a higher level of social and cognitive capacities: '*Cultural learning* refers to a more sophisticated subclass of social learning abilities in which individuals seek to acquire information from others, often by making inferences about their preferences, goals, beliefs, or strategies and/or by copying their actions or motor patterns' (Henrich, 2015, p. 13, original emphasis). While she agrees that there needs to be some way of distinguishing cultural learning among social learning, Cecilia Heyes (2018, p. 85) criticises Henrich's characterisation for not being clear enough to set cultural learning apart from other kinds of social learning, either conceptually or empirically. In particular, Heyes sees an important problem with Henrich's decree that 'When discussing humans, I'll generally refer to *cultural learning*, but with non-humans and our ancient ancestors, I'll call it social learning, since we often aren't sure if their social learning includes any actual cultural learning' (Henrich, 2015, p. 13, original emphasis). Henrich is non-committal on the issue whether non-human animals could have cultural learning, but the distinction he makes here is indeed highly problematic. If cultural learning is in an important way different from other forms of social learning, we cannot make this distinction based on species. First of all, while cultural learning mostly seems to be a human phenomenon, it is at least plausible that non-human primates, for example, also engage in some type of cultural learning (see, e.g., de Waal, 2017). Here I cannot engage further in that topic, but it is important to note the difference between cultural learning being a particular strength of humans and it being considered an exclusively human capacity. In any case, the more important problem I see in the characterisations of cultural learning by Tomasello and colleagues and Henrich is the strong sense of circularity. Even if it were the case that cultural learning is specifically the kind of social learning that humans engage in, using this as a characterisation of cultural learning is circular and thus not particularly informative. The actually interesting questions concern *how* cultural learning differs from other forms of social learning and *why* – if it is indeed exclusive to humans – it is something that only humans engage in.

If species membership cannot be a decisive criterion, something else is needed. In that spirit, Heyes has suggested the following definition of cultural learning: '"Cultural learning" is a subset of social learning involving cognitive processes that are specialized for cultural evolution – for example, they enhance the fidelity with which information is passed from one agent to another' (Heyes, 2018, p. 86). This definition does not include reference to humans as a species, so it is a step forward from the characterisation of Henrich and Tomasello and colleagues. It avoids a circularity implicit in those accounts: the uniqueness of human cultures is not explained through the uniqueness of cultural learning in humans. However, the definition presented by Heyes is not satisfactory, either, because, taken together with the notion of cumulative cultural evolution, it involves another type of circularity. If the purpose of the notion of 'cultural learning' is to explain how cumulative cultural evolution takes place, then it is not particularly informative to characterise the notion through cognitive processes specialised for cultural evolution. In an earlier passage, Heyes writes that cultural learning 'underwrites a whole new inheritance system: cultural evolution' (Heyes, 2018, p. 769). But in the passage quoted above, cultural learning is distinguished among forms of social learning through it being 'specialised for cultural evolution' (Heyes, 2018, p. 86). Stating that cultural learning makes cultural evolution possible is not informative if cultural learning is *defined* as the kind of learning that enables cultural evolution. This is another circularity that we must get rid of.

However, Heyes further elaborates that 'the distinctions between individual learning, social learning, and cultural learning hinge not on *what* is learned, but on *how* it is learned' (Heyes, 2018, p. 86, original emphasis). This indeed seems like a promising line of thought: if cultural learning is in some way a special subcategory of social learning, it is unlikely to be that due to the kinds of things that can be learned. To be sure, it is likely that there are many skills and knowledge that can only be learned through cultural learning. As I have been arguing in this book, arithmetic is a case in point. But there are also many skills that can be learned in different ways, only some of which can be feasibly considered to be cultural learning. Thus, the correct approach to characterising cultural learning would be to distinguish it from other forms of social learning *as learning*.

The example of Heyes is that of 'turnstile learning', that is, going through the turnstiles that one can find in subway stations and sports stadiums. Learning how to go through the turnstiles can take place by watching people insert their tickets and seeing the turnstile open. That is social learning. But what turns it into *cultural* learning, according to Heyes, is when this is done on the basis of verbal instruction: 'In this case, the learning is assisted by another agent, making it social learning, and involves a mechanism – language – that is specialised for cultural inheritance' (Heyes, 2018, p. 87). Again, I am troubled by the potential circularity. If language is seen as a mechanism specialised for cultural inheritance, the characterisation of cultural learning would again be tied to cultural evolution, the very thing it is meant to explain. However, the circularity can be avoided when discussing language. It is problematic to think of language as 'specialised for cultural inheritance', but this does not imply that language could not be highly important for cultural inheritance nevertheless. Indeed, when Heyes in the earlier quote stated that cultural learning processes 'enhance the fidelity with which information is passed from one agent to another', I believe she captured something essential. What distinguishes cultural learning from other forms of social learning, I submit, is that cultural learning takes place through processes of passing information that are *clearer*, *less ambiguous* and *more enduring*.

By 'clearer', I mean that the information is passed with greater fidelity; that individuals can acquire the information without excessive 'noise' compromising the passing of information. By 'less ambiguous', I mean that individuals can acquire the information without an excessive chance of misinterpreting it. By 'more enduring', I mean that the information is not excessively tied to particular events of passing information. All these characterisations are presented in the comparative for a good reason. Information passed on in cultural learning is rarely perfectly clear, unambiguous and enduring. But in cultural learning processes, I propose, these characteristics are enhanced over other types of processes of social learning. As the result, cultural learning involves less inter-personal variability. The information passed across collectives is more un-altered than in other forms of social learning, making it possible for members of a culture to share new practices, skills and knowledge more widely.[2]

Here number concepts and arithmetic provide a perfect example. Why does the Pirahã culture not engage in learning of number concepts, even though they clearly engage in the cultural learning of many other skills, like hunting? The answer lies in that they do not have the cognitive *tools* – or in Heyes' (2018) characterisation, cognitive *gadgets* – for that. Since they do not possess numeral words or symbols, they do not have a sufficiently clear, unambiguous and enduring process for passing

[2] This does not mean that cultural learning always leads to members of the culture widely sharing the trait. Boat-building, for example, is a good example of a skill which is made possible by cultural learning, yet only few individuals in a culture are able to build a boat. While practices, skills and knowledge can spread widely within and across cultures, they can still be learned only by a small subgroup of the culture.

information about number concepts. Even if some members of their culture did acquire number concepts, they would not have the means of communicating them to other members.[3]

Conversely, the introduction of numeral words and symbols can now explain how cumulative cultural evolution of number concepts and arithmetic was possible. Body part counting and tallying are already culturally evolved practices that can be shared among cultures in an unambiguous way, as long as the cultures have the necessary linguistic means, that is, numeral words or symbols. Numeral symbols and writing enabled the passing of information about arithmetic to future generations. These developments made knowledge and skills concerning arithmetic endure. When combined with educational practices, they made those skills increasingly widely spread in societies, up to the modern level in which most members of contemporary societies possess them.

In this development, a great variety of cognitive tools have played a role: fingers and other body parts, tallying objects, numeral words and symbols, writing material, etc. What each of these developments enabled was the passing of information that was increasingly clear, increasingly unambiguous and increasingly enduring. This enabled that we, as part of a culture, *share* our number concepts and arithmetical knowledge. Since we have learned it ultimately from the same materials, we can trust that we have learned them – at least for the most part – in a similar manner. In forms of social learning where information is passed with less fidelity, this may not be the case.

But why is cultural learning something that humans – at least for the most part – have specialised in? How do our cognitive capacities allow for extensive cultural learning and why, for example, do non-human primates, to the best of our knowledge, not possess similar capacities? Some researchers have argued that cultural learning is made possible by genetic adaptations, innate modules and instincts that have developed through processes of biological evolution (Confer et al., 2010). Heyes, however, disagrees with that. According to her, results from comparative psychology, developmental psychology and cognitive neuroscience show 'little evidence that cultural learning is based on cognitive mechanisms that have been genetically adapted specifically to enable the social transmission of information'

[3] This has indeed been shown to be the case with bi-lingual members of the Munduruku, for example, who possess number concepts and arithmetical skills (Everett, 2017).

(Heyes, 2012, p. 2181). Instead, she argues that cultural learning *itself* is primarily a cultural development:

> . . . I propose that the specialized features of cultural learning – the features that make cultural learning especially good at enabling the social transmission of information – are acquired in the course of development through social interaction. This implies that the cognitive processes that comprise cultural learning are themselves culturally inherited; they are cultural adaptations. They are products as well as producers of cultural evolution. (Heyes, 2012, p. 2182)

Here I agree with Heyes, although it is also possible that humans have evolved biologically to have more powerful forms of behavioural and developmental plasticity, as argued by Sterelny (2003, 2012). Even if this were the case, it does not imply that humans have evolved biologically to be unique in the way we can apply cultural learning. It seems more plausible that the methods of cultural learning themselves are 'cognitive gadgets', culturally developed practices and contents that enhance the way we can learn.

Importantly, in addition to obvious culturally shaped creations like languages, these cognitive gadgets also include 'mechanisms of thought' (Heyes, 2018, p. 1). This is consistent with the account developed in this book. Number concepts and arithmetic did not evolve culturally only because of numeral words, symbols and other cognitive tools. They also required a certain numeral and logical way of *thinking*. This way of thinking in terms of numerosities proved to be advantageous for many cultures and they evolved it further, with the help of further cognitive gadgets. It would be misleading to say that non-arithmetical cultures *lack* a way of communicating arithmetical information, given that they do not think arithmetically in the first place. At the same time, they possess many ways of thinking highly useful in their cultures, ones that we in industrialised cultures do not have. This is the essence of cultural learning and cumulative cultural evolution: the methods and contents involved in learning are shaped by the environment and customs of the particular culture.[4]

6.3 Creativity and Innovation

The kind of development of culturally evolved abilities described in Section 6.2 is made possible by a feedback loop in which new innovations

[4] This is an important observation when debunking the dangerous idea that some cultures are in some general way *superior* to others.

make new skills, artefacts and knowledge possible, which then allow for yet further innovations. This brings us to the final stage in understanding the phenomenon of cumulative cultural evolution. Cumulative cultural evolution is possible through cultural learning, which in turn is possible through the clearer, less ambiguous and more enduring passing of information. But for cultural evolution to take place, this body of information, and the methods of passing it, need to change and grow. We know that biological evolution takes place through genetic mutations which prove to be advantageous in processes of natural selection. What is the counterpart of mutations in processes of cumulative cultural evolution, that is, how do cultures create new innovations that are then tested for their advantageousness?

When it comes to the cultural development of arithmetic, it is clear that a great deal of innovation has taken place in terms of – among other features – language, notations, tools and practices. Human and non-human innovation and creativity come in many forms, but in the case of arithmetic we are interested in innovation that has transformed *cognitive* practices. In the broadest sense, every human interaction with the local environment can be seen as being creative: it brings into existence something that did not exist previously. Digging a new ditch, for example, is a creative process in this sense. However, it is not necessarily an innovative process – although it clearly can also be that if, for example, a new way of digging or a new type of ditch is introduced. Similarly, cognitive processes in a broad sense are always creative. Any new thought of an individual creates something new to the world, as does the act of communicating that thought. However, only a small subclass of cognitive processes are innovative in the sense that they can transform cognitive practices wider within a culture.

Due to the potential excessive generality of the notion of creativity, here I will focus on innovation. In particular, my focus is on what Fabry (2017) has called *cognitive innovation*. This notion, following, for example, Lane (2016), takes innovation to concern innovative *processes* instead of innovative products: 'Innovative cognitive processes are characterized by the realization of cognitive procedures that complement, augment, or transform the overall cognitive potential of a certain social group of organisms' (Fabry, 2017, p. 377). Cultural innovation is then best understood as a special case of cognitive innovation, which is present also in non-human animals: '[A]nimal innovation [is] defined as the devising of a novel solution to a problem, or a new way of exploiting the environment' (Laland, 2017, p. 100). The end product of innovative processes can take many forms, from artefacts to patterns of behaviour and belief systems (Fabry, 2017, p. 377). This notion of cognitive innovation would appear to be a good fit with the present account of cultural learning and cumulative cultural evolution of arithmetic. Arithmetical practices include artefacts (pen and paper, abacus, etc.) but also non-physical innovations, such as number words and problem-solving algorithms.

Here I understand cultural cognitive innovation as a special form of cognitive innovation, just like cultural learning is understood as a special form of social learning. In cultural cognitive innovation, information is passed in a clearer, more unambiguous and more enduring way than in cognitive innovation in general. This is possible through the use of language, which itself is in a constant process of changing due to cultural cognitive innovation (as in the case of introducing numeral words). I take this notion of cultural cognitive innovation to be uncontroversial. However, what is less obvious is the placement of agency in cultural innovation. As detailed by Muthukrishna and Henrich (2016), innovation has traditionally often been seen as the work of capable individuals. From Prometheus and Shaka Zulu to Gutenberg and Newton, Muthukrishna and Henrich argue, there is a common narrative of exceptional humans being responsible for the greatest innovations (Muthukrishna & Henrich, 2016, pp. 1–2).[5] Recently, this idea has been challenged by many researchers who emphasise the collective nature of innovation:

> Intelligence is not irrelevant of course, but what singles out our species is an ability to pool our insights and knowledge, and build on each other's solutions. New technology has little to do with a lone inventor figuring out a problem on their own; virtually all innovation is a reworking or refinement of existing technology. (Laland, 2017, p. 7)

Muthukrishna and Henrich (2016) also stress how innovation is made possible by cultural learning abilities which are used within societies. They even go as far as placing innovation in the 'collective brain' of a society of a social network. While I find the ontology involved in this move highly problematic, in general I do agree with the move from individuals to societies or social networks as the principal home of agency in cognitive innovation.[6] However, my reason for this is somewhat different from the

[5] In the vast majority of cases, those narratives have traditionally involved exceptional *men*.

[6] It would be more acceptable to speak of 'collective mind' as a continuation of the frameworks of distributed (Hutchins, 1994) or extended minds (Clark & Chalmers, 1998). It is my opinion that philosophy of mind and cognition becomes needlessly difficult if also physical organisms and their parts, like the brain, are thought of as being extended. As long as the 'collective brain' means nothing

one stated by Laland, as well as Muthukrishna and Henrich. I do believe that exceptional individual efforts are central for innovations. All innovators are standing on the shoulders of giants of past progress, of course, but that does not mean that the individual effort should be under-appreciated.

However, the reason I believe that it is better to focus on collectives is that, on the scale of cumulative cultural evolution, individual innovations cannot be detached from the way they are received by the surrounding social networks. Sometimes individuals can introduce innovations that cannot be described as incremental in the same way as most progress is. But what makes those innovations important is the influence they have on the surrounding social networks. It is the culture around the innovators that embrace, proliferate and further develop the innovations. For this reason, in the framework based on cumulative cultural evolution, it would be mistaken to focus on the 'lone genius', even in cases where that might be otherwise justifiable.

I believe that this move of focus to the societal level is helpful for explaining cumulative cultural evolution also in cases where we can trace individual innovations, but it is the *only* way we can understand much of the cultural history of cognitive innovation. In most cases, we have no way of determining what the contribution of particular individuals has been. While important historical arithmetical innovations are associated with particular names (Euclid, Aryabhata, etc.), it is impossible to establish reliably what was innovated by them and what has simply survived in the writings associated with them. For even earlier developments, such as the introduction of numeral words, we cannot even venture an educated guess concerning the individuals involved. In a field of study like the early cumulative cultural evolution of phenomena like numbers and arithmetic, where there is a dearth of reliable historical information concerning individuals, focus on individuals would seem to be drastically misplaced.

Thus, both for reasons of accuracy and relevance, cognitive innovations should be studied at the level of a collective, whether it is a society, social network or a culture. On this level, technologies and practices are improved incrementally and these improvements are expanded gradually across and between cultures. In Section 4.4, we have seen how such innovation can take place through the example of number concepts and

more than a metaphor for cognitive processes extended into societies, this may not be a serious problem, but it is still difficult to see motivation for introducing the notion.

numeral words. Material engagement with objects in our environment can lead to new cognitive practices, such as tallying or body part counting. Such a development, as we have seen, can be explained based on cumulative cultural evolution: through incremental advances new artefacts, skills, practices and knowledge is developed by cultures. Now the question is, how do such innovations take place? What happens in the transition between a collective not having a particular innovation at some time t and then later having it at a later time t'?

Let us see how such innovation can take place through the example of the cognitive practice of tallying. The first stage of such innovation must be an individual trying something new, whether by accident or on purpose. In the example of tallying, this can be an individual making a notch on a piece of wood. This individual may make further notches or perhaps shows the piece of wood to other individuals who may then engage in this practice. The initial purpose of the notches may not have anything to do with numerosities. It could be, for example, purely decorative. Thus, the initial notches on the wood can be an extension of another, non-numerical practice. At some point, perhaps already taking place in another generation, some individual may then use the notches to keep track of, for example, the amount of prey animals they have observed. They may communicate this to other individuals, who can then develop the practice further. Since keeping track of the prey is useful for the collective by giving important information for the next hunt, this practice will be communicated to yet others, giving more members of the collective a chance to enhance the practice.[7]

We could continue with this example, but, given its entirely speculative nature, perhaps it is best to stop here. The point should be clear. Every enhancement to the cognitive practice of tallying is initiated by an individual, but it is spreading that enhancement to others that makes it a cognitive innovation. As knowledge and skills are spread further, they are likely to be improved through the sheer strength in numbers that the collective can offer. And, given that through cultural learning these innovations can be communicated more clearly, more unambiguously and more enduringly to new generations, this process is not tied to a particular group of people and becomes a *cultural* cognitive innovation.

[7] Dawkins (2016) famously proposed the notion of *meme* as the unit of cultural transmission, mirroring gene as the unit of transmission in biological evolution.

Now all that is left to complete the outline of this account of cultural cognitive innovation is to understand how the individual stage of the innovation is possible, that is, how does an individual end up doing something that they have never seen another individual do. This might seem like a futile question, given how much modern societies value the creation of new things, whether in art, sciences, or other fields of activity. Innovating new things is what humans *do*. All these activities, however, are the product of cultural innovation themselves. To avoid circularity, some more fundamental behavioural pattern must be established. This provides few problems, however, given that for decades it has been known that curiosity and exploration are central parts of lives for all higher animals (see, e.g., Berlyne, 1966). Animals in general do not simply follow the behavioural patterns exhibited by their conspecifics. Instead, they engage in play and discovering new activities all the time. Many similarities between the similarity of play in non-human animals and human children have been established (Power, 2013).

I submit that it is ultimately this drive for play, curiosity and exploration that is behind the individual stage in innovation, which can then spread wider to the collective through cultural learning. It is important to stress here that innovation as such is not unique to humans by any means. Animals can also engage in innovation on the level of the collective, as demonstrated, for example, by the famous sweet potato washing monkeys observed in Japan (Kawai, 1965). However, just as cultural learning should be distinguished from social learning in general, cultural innovation needs to be distinguished from the more general social innovation. This does not mean that there is any substantial difference in the origins of new behaviours that are innovated. Humans, just like non-human animals, engage in novel activities because they are by nature curious. The difference is how human cultures can transmit the advantageous results of such curiosity and thus build on them.

6.4 Enculturation and Cumulative Cultural Evolution

To end this chapter, we should stop to consider how Parts I and II of this book are connected. Clearly an integral question concerning the cultural development of knowledge is how the cultural learning of cognitive innovations is possible on the individual level. This is where the two theoretical frameworks presented here – one based on enculturation in ontogeny and the other on cumulative cultural evolution in phylogeny – are connected. The connection is of course quite natural, given that the strength of the enculturation framework lies in precisely the fact it can provide an account of how cultural learning takes place on the level of individual ontogeny. Enculturation and its neural instantiation were discussed in detail in Chapter 3, so, combined with the considerations in Part II, we now have a comprehensive account of the development of arithmetical knowledge that includes both the level of individual ontogeny and the level of phylogeny and cultural history.

In Section 3.5, I presented five empirically testable predictions of the enculturation account for arithmetical cognition. In this section, I want to return to those predictions from a different angle, that of cumulative cultural evolution and the cultural history of arithmetic. Three predictions are particularly important: Predictions 1, 2 and 4. In Section 3.5 they were presented as ontogenetic predictions, but here I examine them from a phylogenetic and culture historical angle.

Prediction 1: Our proto-arithmetical abilities should limit what kind of mathematical abilities can be learned culturally

Just as there are limits to what kind of mathematical abilities can be learned by an individual, there are limits to what kind of mathematical practices can be developed by a culture. It is not possible, for example, to develop mathematical practices that are prohibitively complex for individuals to engage in. More importantly for the present context, human proto-arithmetical abilities also limit the kind of arithmetic that cultures typically develop. As we have seen in this section, arithmetic – at least when it comes to finite numbers and basic operations on them – has developed in convergent ways independently in different cultures, if it has developed at all. Without the proto-arithmetical origins of arithmetic, this would be an unexpected finding. Why have some cultures not developed, for example, cyclical number systems instead of linear ones, as proposed by Rips et al. (2006)? In the present account, there is no mystery to this. The ANS and the OTS, as detailed in Part I of the book, determine (partly) the kind of number concepts that humans can acquire. Similarly, they determine (partly) the kind of number concepts that humans can develop. A cyclical number system where, for instance, after ten numbers you start over from one, does not conform to this. Of course, such number systems *can* be developed – and have been developed, at least in Western mathematics – but nowhere have such systems been the historically first number systems. The most likely explanation for this is that proto-arithmetical abilities – being universal to humans – constrain the development of number systems and arithmetic in similar ways everywhere in human cultures.

Prediction 2: The difficulty of learning a mathematical practice or skill should depend on the distance between the initial function of the proto-mathematical neuronal circuits and the new ones

Seen from a phylogenetic and cultural historical angle, this prediction implies that, in addition to learning, also *developing* mathematical practices and skills depends on the distance between them and proto-mathematical abilities. This is likely to be the case generally when it comes to mathematics, but here it is not possible to discuss the general question. In the case of arithmetic, however, this certainly appears to be the case. As we have seen, arithmetical practices and skills developed by different cultures converge when it comes to basic operations (addition, subtraction, multiplication, division). This is most likely due to their close relation to counting processes. As detailed in Section 5.1, addition can be learned as a direct continuation of counting, multiplication as a direct continuation of addition and so forth. Whenever in history of arithmetic there have been steps that have required innovation that goes beyond this kind of continuation, we see that these steps are not taken universally. The introduction of zero, infinity and proofs are all examples of this. They are further from the proto-arithmetical abilities than the basic operations are; therefore, it is to be expected that they are more difficult for cultures to develop. While there can be other reasons for not developing practices and skills – e.g., lack of applicability – it is feasible that difficulty is at least one factor.

Prediction 3: Cultural learning of mathematical abilities can reduce the cortical resources available for proto-mathematical abilities

This prediction remains unchanged from the point of view of phylogeny and cultural history, since the cortical resources available for proto-arithmetical abilities on the level of populations depend entirely on processes of enculturation in individuals. Arithmetic is likely to be too late a development to have caused significant changes in the genome of humans in arithmetical cultures.[8] The human cognitive capacity for arithmetic, as far as we know, is the same universally. Of this, however, there are no large-scale studies. Indeed, such a study would be both practically and ethically problematic. It would be extremely difficult to systematically study the capacity of humans with different cultural origins to learn arithmetic under controlled settings. But even if this were the case, would we want to have such a study? If members of some culture did indeed have some kind of innate deficiency for arithmetic, this would be a potentially dangerous result that could cause discrimination. From what we know, however, there is no reason to think that this would be the case. Thus, this prediction is likely to come back to the level of the individual, as described in Section 3.5.

Prediction 4: We should expect increasing inter-cultural differences in mathematical abilities when they are increasingly distant from proto-arithmetical abilities

On the ontogenetic level, this prediction concerned individual abilities in different cultures, but in the context of phylogeny and cultural history it concerns mathematical knowledge and skills. In Section 3.5, I promised to treat this prediction in detail in Part II. Now that this has been done, it should be clear how strongly this prediction is corroborated by anthropological data. Arithmetic has been developed several times independently by cultures, but there have been important cultural differences (Ifrah, 1998). As we have seen, the further we go from natural number systems and arithmetical operations, we can see more such cultural differences. The notion of mathematical proof is a good example of this. Of course, when we expand our approach beyond arithmetic, this difference becomes even more pronounced. To give just one example, in Ancient Greece the existence of irrational numbers was already known, but they were not accepted as mathematical objects (Merzbach & Boyer, 2011). In modern mathematics, irrational numbers are accepted as part of mathematical discourse just as rational numbers are. It is not possible to give a detailed presentation of the topic here, but the history of mathematics follows this prediction

[8] Not everybody agrees: Lumsden and Wilson (1981) famously – and controversially – argued for the 'thousand-year rule', according to which already fifty generations may be enough for significant genetic changes. If that were the case, it is possible that arithmetic may have had some genetic effects, considering that the emergence of number symbols dates back to at least 5,000 years ago. However, such knowledge and skills were most likely for a long time limited to very small parts of the population, which would make it impossible for them to have population-level effects.

everywhere. The further mathematics develops from the proto-arithmetical origins, the more diversity we start seeing. The Mayan mathematician and a contemporary Western mathematician, for example, could agree on the part of arithmetic that concerns operations on finite natural numbers. As seen in this part of this book, this is because basic arithmetic cannot develop in conflict with our proto-arithmetical abilities. The notion of proof, for example, is something different. As mathematical cultures develop, their mathematics becomes increasingly specific to the particular culture, consistent with Prediction 4.

Prediction 5: We should expect there to be mathematical abilities that are more difficult to acquire later in ontogeny, due to cortical resources already having been used for earlier abilities

This prediction concerns the ontogenetic level and is practically unchanged. The only tweak that needs to be made is that, on the level of phylogeny and cultural history, the prediction concerns individuals in a culture. But, given that this is a simple generalisation of the original prediction for the cultural domain, the content remains the same.

In Part I, I focused on the acquisition of number concepts and arithmetical knowledge and skills in the individual ontogeny. In this Part II, I have focused on the way arithmetical knowledge and skills develop on the level of phylogeny and cultural history. We have seen that the present theoretical framework incorporating both enculturation and cumulative cultural evolution provides a coherent basis for epistemology of arithmetic. Arithmetical knowledge in individuals is the result of enculturation. It is based on proto-arithmetical abilities, but acquiring arithmetical knowledge and skills requires learning in a cultural setting. Such a cultural setting for arithmetic is possible because it has developed in processes of cumulative cultural evolution.

This completes the first purpose of the book: to explain how arithmetical knowledge can be acquired and developed. Next, moving more into the realm of traditional philosophy of mathematics, we need to ask what arithmetical knowledge according to this account is *like*. This will be the topic of Part III of this book.

6.5 Summary

In this chapter, I discussed the way knowledge and skills evolve culturally. Starting from the notion of *cumulative cultural evolution*, I showed how the emergence of arithmetical knowledge and skills can be understood as the product of trans-generational cultural evolution. Cumulative cultural evolution, however, requires a particular type of social learning, called *cultural learning* in the literature. To make sense of this problematic notion, I proposed the non-circular characterisation of cultural learning as learning that takes place through processes of passing information that are clearer, less ambiguous and more enduring than other forms of social learning. I also presented an account of how innovation of new contents is possible in the framework of cultural learning and cumulative cultural evolution. Finally, I evaluated the five predictions made in Chapter 3 from the perspective of phylogeny and cultural history.

PART III

Epistemology and Ontology

CHAPTER 7

Conventionalism and Inter-subjectivity

As was discussed in the Introduction, Kant (1787) saw mathematical knowledge – including arithmetic – as being synthetic a priori in character. While there are important contrasts, this epistemological position resembles that of Plato, according to which mathematical knowledge is acquired purely by reason and recollection (Plato, 1992, 127a–b). Modern platonist accounts have developed in different directions, but they share the epistemological tenet that mathematical knowledge is acquired a priori. The problem with such epistemological views, however, is that they appear to demand an epistemological faculty unlike any other we currently accept as part of the standard empiricist–scientific worldview. This problem is most famous in the formulation of Benacerraf (1973), who asked how we can fit a Goldman-type (Goldman, 1967) causal theory of knowing with platonist ontology.[1] If we accept that mathematical objects are abstract and exist mind-independently, in order to get knowledge of them, we need to have some epistemic access to them. But, given that we are physical subjects, how can we get epistemic access to non-physical, causally inactive objects? This, of course, was what Parts I and II of this book were ultimately about, and in this Part III I will spell out the answer to this epistemological question in detail.

I contend that the way the present account deals with Benacerraf's problem is one of its greatest epistemological strengths.[2] In traditional philosophy of mathematics, some of the more famous suggestions for epistemic access to abstract objects include Gödel's (1983) suggestion of a special epistemic faculty for mathematics and Penrose's (1989) mathematical 'seeing'. Another approach has been to deny that mathematical

[1] This was the way Benacerraf presented the problem, but it is not restricted to causal accounts of knowledge (see Field, 1989).

[2] The account in this book is limited to arithmetic, but similar epistemological strength is present also in other accounts based on proto-mathematical abilities, such as the one presented for geometry in Pantsar (2021d).

knowledge is a priori, as we saw in the introduction in the case of Mill (1843), and more recently argued by Kitcher (1983). But, in the modern philosophical literature, such solutions have not found much popularity. Losing the a priori character of mathematical knowledge is widely seen as problematic, but it seems equally unacceptable to postulate special epistemic faculties whose existence is unsupported by any evidence.

This has led to a situation in which some key assumption of epistemology in the tradition of Plato and Kant needs to be given up. For Carnap (1937), the principle to be abandoned was the view that mathematics is *synthetic* a priori, and this position has often been taken in the modern literature (e.g., Linnebo, 2018b; Rayo, 2015). According to this view, mathematics consists of a priori knowledge, but it is always *analytic* a priori. However, if mathematics is not synthetic a priori and as such does not give genuinely new knowledge of the world, then what is mathematical knowledge about? How do we account for the pursuit of knowledge that does not provide genuinely *new* knowledge? Carnap's response was that mathematical knowledge is about *conventions*.

In this chapter, I will argue against conventionalist theories of arithmetical knowledge. When surveying the modern literature on the philosophy of mathematics, it may seem strange that I give conventionalism so much focus. Aside from some exceptions, most notably the recent book by Jared Warren (2020), conventionalist accounts of mathematics have not been popular in the contemporary scene. However, there is a good reason for me to focus on the conventionalist threat, and that is the danger that the epistemological account I have developed is interpreted in a conventionalist way. After all, I see an integral role for culture in the development and acquisition of arithmetical knowledge and culturally developed practices are often understood as conventions.

As I will argue in this third part of the book, I do not believe that arithmetical truths are conventional in the strong sense required by conventionalism. This strong sense means that arithmetical truths are fundamentally *only* cultural conventions. I have called this kind of view *strict* conventionalism (Pantsar, 2021c) and I hold it to be the philosophically most interesting conventionalist position. After all, it is trivial that there is an important conventional element in arithmetic, just as there is in all mathematics. But we should not confuse the existence of the conventional element with arithmetical truths being fundamentally conventions. Moderate conventionalist views – in which I include, for example, Cole (2013, 2015) – state that mathematics (including arithmetic) consists of conventions, but those conventions are not arbitrary. However, in my view that kind of moderate conventionalist position is only interesting if it includes some kind of explanation *why* the conventions are not arbitrary. But if we have such an explanation, how conventionalist is the view anymore? Indeed, in the present account, I believe that there is such an explanation, and I certainly do not see my account as conventionalist.

Hence the philosophically more interesting conventionalist position states that arithmetical truths are conventions and nothing more, that is, they could ultimately be arbitrary. It is this strict conventionalist position that I will mainly focus on in this chapter, and I call the challenge provided by it the 'conventionalist threat'. The conventionalist threat, in other words, poses the problem that if mathematical knowledge is ultimately tied to nothing but conventions, how can we explain why some mathematical statements are accepted as truths and others rejected as falsehoods?

While I believe that the present account survives the conventionalist threat, I do recognise it as a genuine problem and hence worthy of close attention. Indeed, I generally see conventionalism as both an initially appealing position and a serious threat for a feasible epistemology of mathematics. It is appealing because all the traditional competing theories are problematic in ways that conventionalism seems to avoid. Postulating mind-independent mathematical objects is a particularly problematic part of the platonist tradition in philosophy of mathematics. If we take Benacerraf's problem seriously, we need to identify a plausible access-relation between human subjects and mathematical objects. But if mathematical objects are simply postulates based on human conventions, such a problem never arises. Conventionalism has an equally good answer ready against the problems of empiricism and a posteriori views of mathematics: mathematical knowledge is not empirical because it ultimately concerns only human conventions. As such, it is firmly in the realm of a priori.

Saving the apriority of mathematics and not postulating mind-independent mathematical objects are both features that I want to include in my epistemological account, yet I do not want the present account to be interpreted in a strict conventionalist fashion. Admittedly, given that arithmetic has historically had great variety in different cultures, that may seem like a feasible interpretation of my account. In this chapter, however, I will debunk such conventionalist interpretation. I accept that there are important conventionalist elements to arithmetic, ranging from numeral word and symbol systems to arithmetical cognitive tools and applications. However, more importantly for the present matter, the shared proto-arithmetical origins determine that arithmetic develops in largely convergent ways whenever it is developed within a culture. This,

I argue, saves arithmetic from the conventionalist threat. In Chapter 8, I will then show that arithmetical knowledge thus understood actually retains in strong forms all the characteristics traditionally associated with arithmetical knowledge, that is, apriority, objectivity, necessity and universality.

7.1 The Conventionalist Threat

The heyday of mathematical conventionalism was in the early twentieth century when it took its place as the standard philosophy of mathematics for logical positivists. Their view was that mathematical statements are necessarily true, but that is due to them being analytic truths like the statement 'all bachelors are unmarried', devoid of all factual content (Carnap, 1937; Hempel, 1945). Mathematical statements do not state anything about the world, and that is why mathematical knowledge is universal, necessary and a priori in character. To distinguish conventionalism from platonism, it was important that mathematical statements got their content only from human conventions. According to the conventionalist view, mathematics is not about the physical world, but it is not about any mind-independent non-physical world, either. In the words of Ayer, 'The principles of logic and mathematics are true universally simply because we never allow them to be anything else' (Ayer, 1970, p. 77). In short, we have agreed upon mathematical conventions and they are so firmly entrenched that they cannot be changed.

As I understand it, conventionalism was also the view supported by Hartry Field in his book *Science without Numbers*, even though it was presented in terms of nominalism, that is, the view that abstract mathematical objects do not exist (Field, 1980). As a counter-reaction to logical positivism, Quine (1966) and Putnam (1979) developed their position according to which we cannot separate mathematical and non-mathematical scientific content. Mathematical statements featured in scientific theories are true just as other scientific statements are, and the entities they quantify over exist just like the physical entities postulated in the theories do. This may not always be the case, however, since sometimes it is possible to remove mathematics from scientific theories. But, if mathematics is *indispensable* for scientific theory – that is, we cannot remove the mathematical part without also losing some of the non-mathematical part of the theory – we cannot distinguish between physical and mathematical ontological and epistemological commitments (Colyvan, 2001).

Field accepted this *indispensability* argument of Quine and Putnam as the strongest argument for mathematical realism. Therefore, he aimed to show that mathematics is in fact *not* indispensable in science. His treatment focused on Newtonian mechanics and it is generally seen as controversial but, if valid, it would show – or so Field's argument went – that mathematical statements are simply conventions and state nothing about the world.[3] After all, according to him they could be removed from scientific explanations, thus rendering the strongest argument for mathematical realism powerless (Field, 1980). Field later called his position *fictionalism*, a term he used to refer to a particular type of nominalism. The central idea of fictionalism was that mathematical objects do not really exist, but they can exist as part of a fiction, analogously to Sherlock Holmes existing in the fiction of Arthur Conan Doyle's books (Field, 2022). As I see it, fictionalism is comparable to strict forms of conventionalism: if mathematics is analogous to fiction, surely mathematical truths cannot be anything more than conventions.[4]

After the initial attention around Field's book faded away, conventionalism again became an unpopular position in philosophy of mathematics, mainly confined to historical treatments. However, as mentioned, Warren has recently presented a book-length defence of conventionalism (Warren, 2020). Since I consider his formulation of conventionalism to be the one with the most potential, I will focus on it here. One reason for this is that, as he explains (pp. 5–6), the conventionalist view of the logical positivists was hardly uniform. At best, we can say that the different views had a family resemblance in that they made logic and mathematics in some important way to be about human linguistic conventions. To make the matter more precise, Warren distinguishes between three versions of conventionalism. The first of these is *descriptive* conventionalism, according to which logic and mathematics are about the use of logical and mathematical words. Given that this violates the idea that logic and mathematics are a priori – after all, the descriptive project to describe the use has to be a posteriori – Warren (p. 7) dismisses this version. The second version is *non-cognitivist* conventionalism, according to which logic and mathematics are about linguistic rules. However, this

[3] One of the main controversies concerning Field's book is that, while it may be science without *numbers*, as the title says, many mathematicians contend that it is not science without *mathematics* (Shapiro, 2000, p. 232). In this way, Field's program is thought to mirror Newton's mathematical structure of space and time, which would mean that it simply replaces one mathematical domain with another.

[4] This also seems to be the current view of Field (2022). Conventions can be more than fiction, of course, so the relation between fiction and conventions is not symmetric. If fictionalism is correct, mathematics must be conventional in a strict sense.

conflicts with the idea that mathematical statements can be *true*. Given that rules do not have truth values, Warren (pp. 8–9) also dismisses this version. Thus, the relation between logic/mathematics and linguistic conventions can be neither descriptive nor one of identity. Instead, he argues for the *explanatory reading* of conventionalism, according to which linguistic conventions fully *explain* the truth (and necessity) of logical and mathematical statements (p. 9). The resulting idea of mathematical conventionalism for him is that 'Facts about mathematical truth and falsity in any language are fully explained by the linguistic conventions of that language' (Warren, 2020, p. 10).

It seems to me that this is indeed the notion of conventionalism that we should focus on. The key question is, just like Warren points out, whether mathematical truth (and falsity) is *fully* explained by linguistic conventions. Weak forms of conventionalism are trivially true: clearly there is a lot about mathematics (and indeed all discourse) that involves linguistic conventions, starting from the use of numeral symbols and words. The question is not whether such conventions are an important part of mathematics as a cultural phenomenon. It is whether that is ultimately *all* there is to mathematics. Warren argues that it is, and the relevant conventions are 'syntactic inference rules that are rooted in linguistic behavior' (p. 55). Now the key metasemantic question becomes how inference rules rooted in linguistic behaviour come to be established. For Warren this is simply due to 'the basic rules of language that the relevant group of languages users have settled upon' (pp. 103–104). In the case of mathematics, this means that mathematical truths are 'true as trivial by-products of the conventions governing our mathematical language' (p. 202).

Warren's treatment of conventionalism is at its strongest when it focuses on logical conventionalism. Mathematical conventionalism complicates things because in mathematics, unlike in logic, we are committed to existence claims. Warren claims, however, that this provides no additional problems. According to him, the existence of mathematical objects like natural numbers is not a 'substantive and controversial fact about our world, but is instead a *trivial by-product* of our using our arithmetical language in the way that we do' (Warren, 2020, p. 213, original emphasis). Therefore, in Warren's treatment, conventionalism is not a nominalist theory but a realist, *platonist* one. Interestingly, Field and Warren end up having contrasting accounts about mathematical objects, even though they seem to share the commitment to conventionalism. Warren's particular brand of realism, however, takes the existence of mathematical objects to be trivial, merely a consequence of using particular linguistic conventions, 'shadows of our syntactic rules' (p. 237). In Section 9.2, we will turn to other such accounts of the trivial existence of numbers, but let us accept for the sake of the argument that the existence of mathematical objects adds no serious problems to the conventionalist account. As Warren (p. 237) develops the account, the existence of numbers is simply a trivial consequence of us understanding and accepting the rules of Peano arithmetic.

But what is the status of mathematical truths? They are conventions, but are they *arbitrary* conventions? For Warren, this ultimately seems to be a non-question. In one way, mathematical truths as by-products of syntactic rules of our mathematical languages are not arbitrary, because we cannot arbitrarily decide what the syntactic rules are. But he also emphasises that mathematical truths 'have no deeper nature, and they answer to nothing external save for our varying needs – communicative, scientific, and the likes' (Warren, 2020, p. 317). In this way, it seems quite possible that mathematical truths are ultimately arbitrary, given that Warren does not provide any reason why the syntactic rules of our languages could not have developed fundamentally differently. I believe that, ironically, this is the great strength of his treatment of conventionalism: as the strongest conventionalist account in the literature, it manages to pinpoint the key problem of mathematical conventionalism. This problem is that conventionalism ultimately makes mathematical truth contingent on the kind of rules of syntax that our mathematical languages have.

What is the problem, the conventionalist may ask? If mathematical truth is about conventions, what more can we hope from a philosophical account of mathematical truth? Indeed, if we had no other knowledge relevant to the question of mathematical truth, it could be the case that epistemological accounts of mathematics can focus exclusively on conventions. But, as I have shown in this book, there is a wealth of knowledge concerning why arithmetical truths are the way they are, based on our evolutionarily developed, *pre-linguistic*, cognitive capacities. It is indicative of Warren's conventionalist account that none of that is mentioned in the 375-page book with even a single word. He is happy with the position that linguistic conventions are the foundation of mathematical knowledge. But we know better, as has been established in this book. In the case of arithmetic, at least, part of the foundation is clearly pre-linguistic. While these pre-linguistic foundations may well determine linguistic conventions (starting from, e.g., the singular-plural distinction), their existence clearly goes against strict conventionalist accounts. Linguistic conventions do have influence on the development of arithmetic, and I am not claiming that they should not be seen as important for epistemology of arithmetic. Instead, my point is that linguistic conventions are not *all* there is to arithmetic, and we can analyse the nature of arithmetical truth in a non-linguistic context.

The upshot of this is that we can go beyond strict conventionalist accounts, because we can explain *why* arithmetical languages have the syntax they do, at least when it comes to finite arithmetical operations.[5] This is not an arbitrary matter in any understanding of the word, because it is due to our cognitive architecture. And, as appealing as the conventionalist account may sound, in the case of arithmetic the account presented in this book improves upon it. We can do better than justify arithmetical truths based on them being linguistic conventions; we can provide at least some explanation *why* the linguistic conventions are the way they are. And, given that this explanation is non-conventional, the present account provides a strong argument against Warren's conventionalist position, according to which mathematical truth can be *fully* explained by linguistic conventions.

7.2 Wittgenstein on Mathematics

While I believe that Warren's treatment of conventionalism is the strongest modern account, many philosophers are likely to associate mathematical conventionalism with the work of Wittgenstein. Indeed, given how the literature on the epistemology of mathematics has developed into its present form, it is not possible to ignore the philosophy of Wittgenstein in this book. The problem is, as important as Wittgenstein is often seen to be for the philosophy of mathematics, there is little consensus on how his writings on mathematics should be interpreted. This is particularly the case with the so-called 'late' Wittgenstein, as in books like *Philosophical Investigations* (1972), *Lectures on the Foundations of Mathematics* (1976) and *Remarks on the Foundations of Mathematics* (1978). Clearly Wittgenstein was against platonism, that much can be said for certain. He writes, for example, that 'one talks of mathematical discoveries. I shall try again and again to show that what is called a mathematical discovery had much better be called a mathematical invention' (Wittgenstein, 1976, p. 22). But this prompts the question of what Wittgenstein's notion of mathematical invention is, which is a matter of controversy among

[5] This is not to suggest that the present approach is limited to finite arithmetical operations. It is plausible that geometry, for example, can be explained in a similar way with a similar approach.

Wittgenstein scholars. There are some that see Wittgenstein as a conventionalist, while others believe that to be thoroughly mistaken. As detailed by Ben-Menahem (1998), the problem is that Wittgenstein's views on conventionalism seem to give rise to a dilemma. On the one hand, Wittgenstein's later work emphasises how everything that is considered to be a necessary truth (as mathematical truths would standardly be) is only so due to the arbitrary rules of a particular grammar: 'The only correlate in language to an intrinsic necessity in an arbitrary rule' (Wittgenstein, 1974, p. 184). On the other hand, in Section 2.5 we saw the rule-following problem associated with Wittgenstein (1972) and in particular Kripke's (1982) presentation of his work. According to this view, rules – presumably including those of grammar – under-determine their conclusions, so we can never establish beyond doubt that a particular case is the consequence of a rule.

Therefore, we arrive at the dilemma. Wittgenstein can be seen as promoting conventionalism, in which mathematical truths arise due to following rules. But, since we cannot follow rules unassailably, this kind of conventionalism would seem to fail. For Dummett (1959), the matter was straightforward. He thought that Wittgenstein was clearly a conventionalist, but he was that in a more radical way than the first horn of the dilemma suggests. Dummett called the position expressed by the first horn 'modified conventionalism', according to which some conventions (e.g., mathematical axioms) are fundamental and the rest follow necessarily from them by following rules. Instead of this notion, Dummett saw Wittgenstein as a proponent of 'full-blown conventionalism', according to which not only each rule, but also each *instance* of applying a rule, is a convention. Under this radical position, it is not the case that the Dedekind–Peano axioms, for example, are agreed-upon conventions and arithmetical theorems follow necessarily from them. Instead, every single arithmetical theorem is a convention on its own. Agreeing that $7 + 5 = 12$, for example, is to accept a new convention, instead of agreeing that it follows from the Dedekind–Peano axioms: its necessity only follows from 'our having expressly decided to treat that very statement as unassailable' (Dummett, 1959, p. 329).

The interpretation Dummett gives of Wittgenstein's philosophy of mathematics would not only make Wittgenstein a strict conventionalist, but even among that classification an exceptionally radical one. Indeed, Dummett's version of Wittgenstein is conventionalist up to the point of absurdity. If every new mathematical statement is accepted by our 'express decision' to treat it as unassailable, the emerging picture goes fundamentally against mathematical practice. Do we need to make an express decision to conclude that 2,348,932 + 8,347,232 = 10,696,164? No. We *do* follow rules, or rather in cases like that, we trust that the calculator or a computer follows an algorithm. There is no express decision made when I calculate that particular sum, even if I happened to be the first person ever to calculate it. The same goes for logical inferences and other types of rule-following. Clearly not every new inference is accepted by an express decision. We follow rules all the time and, if there is a convention that is accepted, it is the convention of following that rule, instead of applying the rule in the particular instance.

Thus, the way Dummett presents Wittgenstein is in practice a *reductio ad absurdum* of Wittgenstein's conventionalism, an invitation to reject his philosophy of mathematics as thoroughly unfeasible. It is no surprise then that Dummett's interpretation is highly controversial. Among the first to criticise it was Stroud (1965), who saw Wittgenstein as expressly denying the full-blown conventionalist view that 'everybody could infer in any way at all' (p. 505). Stroud argues that, while Wittgenstein questions the possibility of *justifying* rule-following procedures, this does not mean that any result could be correct. This is because deviant cases of rule-following procedures would not be seen as rule-following at all; calculating, inferring, etc. follow certain rules because that is what they *mean*. In the case of logical conclusions, Stroud (p. 505) quotes the following passage of Wittgenstein: 'The steps which are not brought in question are logical inferences. But the reason why they are not brought in question is not that they "certainly correspond to the truth" – or something of the sort – no, it is just this that is called "thinking", "speaking", "inferring", "arguing"' (Wittgenstein, 1978, p. 155). Thus, arbitrariness is avoided simply because, by adopting the particular practice, we rule out the possibility of deviant cases to following the rule. While that cannot be justified through non-conventional means, Stroud argues, it does not mean that the practices are any less established:

> Because these procedures cannot be given a 'justification' it does not follow that they are shaky or unreliable, or that we are courting trouble if we decide to engage in them. We do not decide to accept or reject them at all, any more than we decide to be human beings as opposed to trees. To ask whether our human practices or forms of life themselves are 'correct' or 'justified' is to ask whether we are 'correct' or 'justified' in being the sorts of things we are. (Stroud, 1965, p. 518)

Therefore, the dilemma presented earlier in this section would be solved because the real problem involved in the second horn, the rule-following, is not that there can be deviant ways of following the rule, it is trying to look for *justification* for the rule. The rule is a convention and conventions cannot be justified. This does not mean, however, that the rules themselves are any less established. As interpreted by Ben-Menahem: '… Wittgenstein sees the internal relation between the rule and its application as the strongest connection possible. The paradox, or rather, the pseudo-paradox, is a symptom of the pathological drive to transcend the limits of language which Wittgenstein aspires to cure us of' (Ben-Menahem, 1998, p. 119). Now we can see why Stroud and Ben-Menahem think that Dummett's interpretation of Wittgenstein is misguided. The problem Dummett had was that there does not seem to be any way of discriminating between different conventions and, thus, every single inference and calculation would need to be justified. As a result, there would not seem to be any way to save this position from arbitrariness. But, as Stroud and Ben-Menahem both argue, the problem is looking for justification where none can be established. Our practices of inference and calculation are not arbitrary, simply because the conventions we follow are not arbitrary and we do not follow them in an arbitrary way. But there is nothing more we can say about that conclusively, other than the fact that humans are more compelled to follow some conventions rather than others. In the analogy of Stroud:

> Logical necessity, he says, is not like rails that stretch to infinity and compel us always to go in one and only one way; but neither is it the case that we are not compelled at all. Rather, there are the rails we have already travelled, and we can extend them beyond the present point only by depending on those that already exist. In order for the rails to be navigable they must be extended in smooth and natural ways; how they are to be continued is to that extent determined by the route of those rails which are already there. (Stroud, 1965, p. 518)

Full-fledged Wittgenstein-scholarship goes beyond my expertise, so here I will refrain from taking a stand on which interpretation of Wittgenstein – that of Dummett or that shared largely by Stroud and Ben-Menahem – is correct. More important for the present purposes is the complete lack of plausibility of the view Dummett ascribes to Wittgenstein. This kind of full-blown conventionalism makes everything ultimately arbitrary. The problem is, while neither Stroud nor Ben-Menahem believe that Wittgenstein endorsed such radical conventionalism, ultimately neither of them provides a reason why Wittgenstein's position can avoid succumbing to exactly the kind of full-blown conventionalism that Dummett identified.

I do agree with Stroud and Ben-Menahem that it is unlikely that Wittgenstein endorsed full-blown conventionalism. Based on his later writings, his key insight into conventionalism indeed seemed to have been that conventions are followed, not justified. Yet as an argument against arbitrariness, this is inconsequential. It goes without saying that people follow many arbitrary conventions. For example, in most European countries, people nod their heads to express agreement. We may not be able to justify this convention other than acknowledging that it is a convention. In Bulgaria, however, people shake their heads to express agreement and nod to express disagreement. Such conventions would seem to be a very good fit with the Wittgensteinian account as interpreted by Stroud and Ben-Menahem. People follow the 'nodding rule' of their particular culture because it is a convention, and it could be futile to ask for further justification.

Mathematics, however, does not work like that. Cultures differ in terms of what a nod expresses, but there is no culture with the convention $2 + 2 = 5$. This should be considered an extremely significant matter in the question of conventionalism. I do acknowledge that, as individuals, we initially learn to follow the rule $2 + 2 = 4$ because it is an established convention, but to stop the analysis there and not look for further justification cannot be satisfactory. I am prepared to agree with Wittgenstein in that there are conventions that resist justification. However, I do not think that arithmetical conventions are such.

Of course, Wittgenstein did not have access to the kind of understanding about the ontogenetic and phylogenetic development of numerical abilities that we currently do. From his perspective, it may have not made sense to look for justification for rules of mathematics. Whether this should be thought to lead to full-blown conventionalism as suggested by Dummett is controversial. But the problem Dummett presented in interpreting Wittgenstein was valid. If the following of conventions resists justification, it is possible to understand Wittgenstein's view as 'conventions all the way down', ending up in the radical position ascribed to him by Dummett. Stroud and Ben-Menahem argue against this based on Wittgenstein's remarks to the effect that conventions, even if not justifiable, are not necessarily arbitrary. But the real question is *why* they should not be arbitrary. To follow the analogy of rails above, *why* do the existing rails go in one direction instead of another? If there is no way to provide a justification, how can we establish that the convention is not ultimately arbitrary?

How does this problem relate to the historical development of arithmetical knowledge in particular? As I understand it, the threat of arbitrariness is strongly present in Wittgenstein's account. According to

Wittgenstein, we do not accept that $2 + 2 = 5$ because that is not what addition means to us. But, as far as we know, ultimately the matter of accepting or rejecting this convention may or may not be arbitrary. Without further input, we cannot decide. This, however, would be a highly problematic position. Given that no known culture accepts that $2 + 2 = 5$, we should not be happy with even the *possibility* that rejecting it is ultimately just an arbitrary convention.

7.3 Maximal Inter-subjectivity of Arithmetical Knowledge

An important aspect of conventions is that they come in different levels of arbitrariness. Take the conventions of nodding and shaking head to express positive and negative responses. While the association of nod and shake to 'yes' and 'no' seems largely arbitrary, as evidenced by it differing across cultures, the convention of using head movement to express approval or disapproval can be explained in a non-arbitrary fashion. For instance, people are universally likely to encounter situations in which they need to be quiet, making non-verbal expressions of approval and disapproval useful. In addition, it seems likely that humans run into situations where their hands are not free or visible, ruling out hand gestures. Thus, we can detail circumstances in which gestures through head movement clearly have advantages over other forms of communication. In this way, while the meaning of the particular gesture may ultimately be arbitrary, the convention of using head movement gestures is in itself not an arbitrary matter.

Coming back to the topic of numerical and arithmetical cognition, finger counting provides a similar case to head movement gestures. In Japan, for example, people typically count with their fingers in a reverse manner compared to Western Europe. An open palm indicates 'zero' and 'one' is indicated by folding the thumb, 'two' by folding the thumb and the index finger and so on, until 'five' is indicated by a closed fist (Butterworth, 1999). There are many such cultural differences in finger counting and other routines of body part counting, but some kind of finger counting system seems to be universally in place in arithmetical, but also many non-arithmetical, cultures (Butterworth, 1999; Ifrah, 1998). Instead of being simply tools for learning numerals, adults in arithmetical cultures continue to use finger counting routines after they have acquired abstract abilities with natural numbers (Hohol et al., 2018).

Similarly to head movement gestures, then, the particular finger counting routine adopted by a culture seems to be a largely arbitrary matter. However, the adoption of the general convention of finger counting (and other routines of body part counting) can be explained.

As detailed in Section 6.1, finger counting routines help acquire number concepts and arithmetical skills, and finger gnosis levels predict arithmetical learning success in young children (Reeve & Humberstone, 2011). It is also easy to understand how finger counting routines can help arithmetical adults. In situations with a lack of opportunities for writing, fingers can be used to externalise numerical information and thus help with working memory limitations. Perhaps most importantly, both for learning and applying arithmetic, fingers (usually) provide us with instant access to ten discrete objects with a stable order. As seen in Section 5.1, manipulating physical objects is beneficiary for learning arithmetic. Perhaps fingers as objects are not special because they are parts of our bodies, but they are special in that they are objects that are almost always available, at least for most humans.

However, the analysis here has so far ignored perhaps the most important factor in the success of finger counting. Finger counting routines would not be introduced in cultures unless we *shared* the physical structure of our hands with other members of our species. An integral part of social learning is imitation (Tomasello, 1999) and the shared physical structure enables imitating finger counting. It is important to note that, since the routine itself varies across the cultures, the shared physical structure is not enough to guarantee that a configuration of fingers communicates numerosity accurately. After all, a folded thumb and four extended fingers means 'one' in Japan and 'four' in Europe. But, for people enculturated with the same finger counting routine, the configuration of fingers is an accurate representation of numerosity. The finger counting practice is thus *intersubjective* among the members of a particular culture.

We have seen that finger counting routines or other guided manipulations involving physical objects, such as counting games, are important for acquiring number concepts and learning arithmetic. However, they are not mere conventions, even though they differ across cultures. First, as we have seen, we can explain in a non-conventional manner why having a finger counting practice can be useful. But perhaps more importantly, the *structure* of finger counting routines is not conventional. While the Japanese and West European routines differ, they are *isomorphic*. In both, you start with one hand configuration and one movement (folding or extending) indicates moving one step forward on the counting list.[6]

[6] Here I frame this in terms of a counting list, but of course for finger counting it is not necessary to possess numeral words or symbols.

Why is this the case? Why have different cultures developed isomorphic body part counting routines? The answer is two-fold. First, it is because we share our bodily structures, our body parts thus providing similar *affordances* across cultures (Gibson, 1979).[7] Second, it is because we share our proto-arithmetical abilities. The shared bodily structure enables different individuals to follow the same finger counting routines within a culture. The shared proto-arithmetical origins ensure that they are linked to numerosities in a similar manner. Importantly, given that proto-arithmetical abilities are shared universally, we now have an explanation of why different counting systems are isomorphic (at least up to some finite number). Counting systems, whether based on body parts or verbal, are partly constrained by our proto-arithmetical abilities.

This finally allows us to spell out why conventionalism over arithmetic is mistaken. It also shows why we do not need to succumb to Wittgenstein's position that mathematical conventions cannot be further analysed. In Wittgenstein's view, we do not accept deviant sums like $2 + 2 = 5$ because that is not what addition means to us. While there may be some conventions that cannot be analysed further, certainly this type of explanation is unsatisfactory if we can have at least a partial explanation why a particular convention has emerged. And indeed, we are now in a position to explain why addition does *not* mean that to us: it is because $2 + 2 = 5$ would go against our universally shared proto-arithmetical abilities. Much about our arithmetic is about conventions: the symbols, the words, the particular practices and so on. Some of these may be arbitrary and some may not.[8] But, in proto-arithmetical abilities, as well as in shared physical structures, we find origins for arithmetic that are clearly not conventional. The proto-arithmetical abilities and fingers may be products of chance, as all evolutionary adaptations are, but, crucially, they are not *arbitrary*.

The proto-arithmetical abilities and shared physical structure makes arithmetical knowledge highly inter-subjective. This kind of inter-subjectivity goes beyond our finger counting routines or head movement gestures. In Pantsar (2014), I argued that the universally shared proto-arithmetical abilities in fact make arithmetical knowledge *maximally inter-subjective*. 'Maximally' has a particular meaning in this context.

[7] Shortly put, Gibson's (1979) notion of affordance means what kind of possible actions the environment offers to an animal.

[8] With symbols, for example, both can be the case. The Roman numeral symbol III, for example, is clearly not arbitrary because it consists of three strokes. The symbol D, on the other hand, seems to be arbitrary.

My position is not merely that arithmetical knowledge is inter-subjective, but that it is so in a special way. This inter-subjectivity does not mean that arithmetical knowledge is universal, which we know not to be the case. It means that arithmetical knowledge, at least up to basic operations on finite numbers, will always develop in converging ways. There cannot feasibly be an arithmetical convention according to which $2 + 2 = 5$. This is not simply because we would not call such a convention arithmetic, as claimed by Wittgenstein. It is because it would be a convention that goes against our very basic forms of cognition. For knowledge that can only develop in accordance with those basic forms of cognition, as is the case of arithmetic, I believe *maximally* inter-subjective is an accurate description.

7.4 Mathematical Conventions

The view that arithmetical knowledge should be characterised as maximally inter-subjective clearly goes against the conventionalist position, but this should not be interpreted as downplaying the role of conventions in arithmetic. Particular symbols and notation practices, for example, are clearly matters of convention, but the conventional character goes beyond such physical manifestations of arithmetic. Numeral systems, for example, have varied a great deal between cultures, from the sexagesimal (base-60) system used in Mesopotamia to the vigesimal (base-20) system of the Mayans to our current decimal (base-10) system (Ifrah, 1998). In modern arithmetical practice, children mostly learn a decimal system and the Indo-Arabic numeral symbols, but traces of other numeral systems can still be seen in many applications. That a circle is 360 degrees and an hour consists of 60 minutes are remains from a sexagesimal system. That ninety-nine is *quatre-vingt-dix-neuf* (literally, four-twenty-ten-nine) in French is a residue of a vigesimal system and so on.

While these residues of non-decimal systems may be mathematically irrelevant – after all, 360 degrees, for example, can be replaced by 2π in radians – numeral systems also have a substantial influence on their applications. These may reach far beyond the apparent characteristics of different numeral systems. As pointed out by Everett, for example, the threshold of what is considered to be statistically significant appears to be highly influenced by our choice of a certain numeral system (Everett, 2017, p. 255). The generally accepted measure of statistical significance is a p-value of less than 0.05. This means, roughly speaking, that there is a less than a 5 per cent chance that the observed data is consistent with the null hypothesis. The null hypothesis states that there is no statistical significance in the observed data, so a *p*-value of less than 0.05 is seen as a threshold for rejecting the null hypothesis, and hence grounds to accept the alternative hypothesis that there is a statistical significance.

While some scientists have argued for the need for better tools (see, e.g., Rafi & Greenland, 2020), the 0.05 *p*-value (and the stronger 0.01 *p*-value) remain well entrenched as criteria for statistical significance. While standardly accepted, it is important to note that 0.05 here is just an arbitrary number and there is no good reason why it could not be 0.04 or 0.06. While this may not seem like an important difference, in practice it can be the difference between accepting some experiment as confirming a hypothesis or rejecting it. But if the same experiment is run 100 times, a *p*-value of 0.05 means that if the null hypothesis were true, the result would be seen five times. A *p*-value of 0.06 means that the result would be seen six times. How can such a small difference be the criterion for accepting a result as statistically significant?

It is not possible to enter that topic here in detail, but I do want to emphasise the conventional nature of *p*-values. Of course, in a proper use of *p*-values, the different statistical significances are taken into consideration. But way too often one encounters *p*-values as dividing results neatly into categories of statistical significance. This makes little sense when we remember that the most common thresholds of 0.05 and 0.01 are clearly consequences of having a decimal numeral system. For us, five and ten are special numbers, most likely due to the physiology of our hands. But they cannot be special in terms of statistical significance simply because we happen to have five fingers on one hand. Therefore, the exact *p*-value thresholds of 0.05 and 0.01 are clearly conventions. They could equally well be, say 0.04 and 0.008, which could make a difference in what findings are accepted in science. In this way, the conventional aspects of arithmetic can have a much wider significance than we might first think.

While *p*-values are about conventions of how to use mathematical statistical tools, conventions can play a big part in mathematics itself. One convention that we are so familiar with that we rarely recognise it as a convention is that the product of two negative numbers is a positive number. As pointed out by mathematicians (see, e.g., Stewart, 2006), there are good mathematical reasons for this convention. Early on in education, we come to accept that, for example, $(-5) \times (-5) = 25$. But what would be the interpretation of this product in an account like that of Lakoff and Núñez (2000), which sees mathematical operations as metaphorical counterparts of physical operations? Subtraction of a negative number from another negative number can be represented as movement along the number line. For multiplication of two negative numbers, however, there does not seem to be such an immediate physical representation. What could be the understanding of 'negative times something'?

There are many other such examples of conventions in mathematics and most students are familiar with some of them, such as dividing by zero being undefined in standard arithmetic. But few students probably stop to think that the status of these conventions could be something different from the other rules they learn in mathematics. For most students, I conjecture, there is no significant difference in the status of mathematical rules that they learn. I believe that conventionalism in the philosophy of mathematics gets its power from a similar source: since there are mathematical rules that we accept as conventions without being able to provide feasible non-conventionalist explanations, perhaps *all* mathematical rules could ultimately be such conventions. Indeed, I think that this is a feasible assumption to make when it comes to mathematical truths. If there were no evidence to suggest otherwise, it would be plausible that all mathematical truths are true by conventions and, following Wittgenstein, it would be a hopeless pursuit to look for justification for these conventions. However, as we have seen in this book, in the case of arithmetic at least, there is a wealth of evidence pointing to arithmetic being (partly) constrained by our evolutionarily developed proto-arithmetical abilities. Arithmetical conventions come in different forms but, due to the proto-arithmetical origins, we know that there are limits to what kind of conventions can be accepted. The conventions we have in arithmetic could never imply, for example, that $2 + 2 = 5$. This proto-arithmetical foundation makes, I have argued, arithmetic maximally inter-subjective, which is a stronger form of inter-subjectivity than practices based on arbitrary conventions can achieve. In Chapter 8, we will see just how strong this inter-subjectivity is and what consequences it has for the epistemology of arithmetic.

7.5 Summary

In this chapter I have focused on the 'conventionalist threat' against the account presented in this book. I presented Warren's conventionalist account of mathematics and discussed Wittgenstein's philosophy of mathematics and its debated conventionalism. I identified strict conventionalism according to which mathematical truths are fundamentally arbitrary as the main threat against the present account. I then showed that, due to arithmetic's foundations in the evolutionarily developed proto-arithmetical abilities, my account survives the conventionalist threat. Arithmetical knowledge, under the present understanding, is *maximally inter-subjective*, which is enough to deal with the threat of arbitrariness. I then discussed the nature of mathematical conventions further, pointing out their importance without succumbing to the view that arithmetic is fundamentally based only on conventions.

CHAPTER 8

The Character of Arithmetical Knowledge

In Chapter 7, I have argued that arithmetical knowledge is maximally inter-subjective. We have seen that this is inconsistent with conventionalism, because the content of basic arithmetic (i.e., arithmetical operations on finite numbers) is shaped by our evolutionarily developed proto-arithmetical abilities. Clearly we do not get to decide what statements are arithmetical truths, but neither is the reason behind the truth of arithmetical statements 'simply because we never allow them to be anything else' (Ayer, 1970, p. 77). Arithmetical truth is stronger than that. Even if we wanted to allow arithmetical truths to be something else, we could not. This, of course, has been used to great effect in literature. Both Dostoevsky (1864) in *Notes from Underground* and Orwell (1961) in *Nineteen Eighty-Four* made important use of the flawed equation $2 + 2 = 5$ in their novels. For Dostoevsky, believing in it was a mark of human freedom from the oppression of logic and reason. In Orwell's novel, the equation was used as a vehicle of oppression in a dystopia where a political system could distort even arithmetic. In both cases, the flawed equation is a powerful literary tool because it is so unimaginable for readers. If we accept $2 + 2 = 5$, it is something truly extraordinary. Based on all that we have been through in this book, we now have a better understanding of why it would be so extraordinary. It is not because a particular system of reasoning is forced upon us. It is because, ultimately, our proto-arithmetical abilities do not allow us to think otherwise.

However, even though I reject conventionalism, I do not subscribe to any kind of platonist philosophy either.[1] Thus, in the present account we face important questions concerning the nature of arithmetical knowledge. Perhaps the most immediate of these is whether arithmetical knowledge is

[1] Although, as we have seen in Chapter 7, there are also platonist forms of conventionalism, like that of Warren (2020). Stronger forms of platonism, according to which mathematical truths are independent of human conventions, are incompatible with conventionalism.

objective. I have argued that arithmetical knowledge is maximally inter-subjective, but what is the relation between inter-subjectivity and objectivity? How about *universality*, another characteristic that is often connected with mathematical knowledge? Returning to the considerations on Kant in the introduction, we also need to ask whether arithmetical knowledge understood in the present manner is a priori or a posteriori. And, finally, mathematical knowledge has traditionally been understood to be *necessary*, to distinguish it from contingent forms of knowledge.

In this chapter, I present answers to all these questions. As we will see, we need to adjust our understanding of some of the notions under consideration. The traditional interpretations of those notions, suited to views like platonism and conventionalism, will not fit the present account. However, I argue that, under a proper understanding of the terms, arithmetical knowledge can be feasibly understood as being a priori, objective, universal and necessary.[2]

8.1 Apriority

In Kant's view, mathematical knowledge is characterised as synthetic a priori, which I also take to be the mainstream traditional interpretation of platonism (see, e.g., Linnebo, 2018a). In conventionalist accounts, as we saw in the last chapter, mathematical knowledge is usually seen as being analytic a priori. Here I will not focus on this issue of analyticity vs. syntheticity, although I will return to it later. As has become clear, I do not think that mathematical knowledge concerns only linguistic conventions. However, in the present account, it is still possible that mathematical knowledge is analytic, true only in virtue of the meaning of its concepts. The difference to conventionalism that I have stressed is that the meaning of those concepts does not come only from conventions, but also from our evolutionarily developed cognitive architecture.

This is not to suggest that the question of analyticity and syntheticity is not a *bona fide* philosophical problem. Quine (1951) famously argued that the distinction between synthetic and analytic is a dogma of empiricism which can only be defined in a circular fashion. But, like Putnam (1983), I think this is an over-simplification. I do think that there is a fruitful distinction to be made between synthetic and analytic truths, and truths like 'all bachelors are unmarried' deserve a special semantic status.

[2] To be precise, the view is that arithmetical *propositions* can be feasibly understood as necessary. Arithmetical knowledge concerns such propositions.

Whether '2 + 1 = 3' has the same status is a well-placed question, and the different answers that Kant and the conventionalists give to it seems like a worthwhile philosophical topic.

However, in the present context, the more interesting epistemological question is whether mathematical knowledge is a priori in the first place, not whether it is synthetic or analytic. The starting point to my treatment of the topic is the observation that arithmetical knowledge certainly has a strong *appearance* of being a priori, understood in the Kantian way as knowledge that can be acquired essentially independently of sensory experience.[3] In Peano arithmetic, for example, the symbol '3' is defined as applying the successor operation S three times to the symbol '0', which denotes the first number. Its successor '$S(0)$' is denoted by '1', whose successor '$S(S(0))$' is denoted by '2', whose successor '$S(S(S(0)))$' is denoted by '3'. Addition as an arithmetical operation is defined so that $x + y$ denotes using S to the number x a total of y times. Thus understood, '2 + 1' simply means applying the successor operation three times to 0, which is exactly what '3' means. Hence '2 + 1 = 3' *prima facie* appears to be an a priori truth. After all, it appears that we could carry out this process without any recourse to sensory experience.

However, we must remember that the Dedekind–Peano axiomatisation, like all other formalisations in mathematics, is a late development following millennia of arithmetical practice and an even longer period of proto-arithmetical activity. In order to be able to carry out the kind of cognitive processes involved in the above construction of $2 + 1 = 3$, one already has to possess number concepts. As we have seen, such concepts are not present in all cultures. They emerge from particular cultural practices in which children are enculturated with new conceptual content and cognitive tools. In these practices, as we have seen, many experiential processes are crucial, starting from finger counting and counting games with physical objects.

Does that suggest that arithmetical knowledge could be a posteriori, empirical in the sense suggested by Mill (1843)? Not at all. Here we must remember the distinction between the context of *discovery* and the context of *justification* (Frege, 1879). Arithmetical knowledge may be discovered by the individual in their ontogeny via empirical observations related to, for

[3] Here the qualifier 'essentially independently' is used because surely *some* sensory experience is needed in order to acquire arithmetical knowledge in a cultural account like the present one, even if it only means, for example, vision used for reading textbooks or the visual and tactile experience related to finger counting.

example, fingers being extended, but of course this is not how arithmetical knowledge is justified. The results of arithmetical operations are not corroborated by carrying out physical manifestations of them, nor is it possible to refute arithmetical theorems by observations. With these characteristics, any interpretation of arithmetical knowledge as a posteriori seems to be misguided. Here I accept that, at the very least, arithmetical knowledge has a strong appearance of being a priori. Nevertheless, I insist that it is a priori in a different sense than a sentence like 'all bachelors are unmarried' is. The reason for this is that the meaning of the notions in that sentence is not constrained by our cognitive architecture in a similar way. 'Bachelor' and 'unmarried' are both conventional terms which have come to, or have been designed to, be related to each other in a particular way. Number concepts, on the other hand, are related to each other in a particular way because they are (partly) based on our proto-arithmetical cognitive capacities.

This distinction is important. When we start learning arithmetic and acquire arithmetical concepts, we do so after extensive experience of treating quantitative information based on our proto-arithmetical abilities. We do not simply learn the meanings of new terms, such as numeral words, from scratch. When we learn their meaning, we adapt and assimilate them to the existing quantitative skills based on proto-arithmetical capacities. Therefore we, as humans enculturated in arithmetical cultures, always learn arithmetic in a *proto-arithmetically set context*. The upshot of this is that, unlike a computer, which could process arithmetic purely based on acquiring an axiom system, for humans arithmetical knowledge is always embedded in a particular cognitive and cultural context. It is in this context, I contend, that arithmetical knowledge is a priori. This is why arithmetical knowledge is best described in epistemological terms as being *contextually a priori* in character (Pantsar, 2014). In the contextual a priori account, not only the acquisition of arithmetical concepts and knowledge takes place in the proto-arithmetically set context. Crucially, the *justification* of arithmetical propositions also takes place in that context. This justification is a priori in character, but it is not independent of the proto-arithmetically set context. We could not justify arithmetical propositions that contradict the proto-arithmetical abilities.

In the philosophical literature, two well-known accounts of contextual a priori knowledge have been presented. Kuhn described in his later writings his famous notion of *paradigm* in terms of being contextually a priori:

> [The notion of paradigm resembles] Kant's a priori when the latter is taken in its second, relativized sense. Both are constitutive of *possible experience* of

> the world, but neither dictates what that experience must be. Rather, they are constitutive of the infinite range of possible experiences that might conceivably occur in the actual world to which they give access. Which of these conceivable experiences occurs in that actual world is something that must be learned, both from everyday experience and from the more systematic and refined experience that characterizes scientific practice. (Kuhn, 1993, pp. 331–332, original emphasis)

Of similar spirit was Putnam's notion of contextual a priori:

> [T]here are statements in science which can only be overthrown by a new theory – sometimes by a revolutionary new theory – and not by observation alone. Such statements have a sort of 'apriority' prior to the invention of the new theory which challenges or replaces them: they are *contextually a priori*. (Putnam, 1983, p. 95, original emphasis)

While both of these accounts resemble my account of contextually a priori knowledge, there are important differences. For Putnam, the example of contextually a priori knowledge was Euclidean geometry, which he held to be a false theory concerning the physical world, yet contextually a priori knowledge. While I do think that Euclidean geometry can be contextually a priori knowledge, I have something slightly different in mind. As I see it, it is more accurate to say that when the general theory of relativity took its place as our preferred theory of macro-level physics, Euclidean geometry stopped being applicable in the new physical theory, rather than it becoming false as a theory of the physical world. Euclidean geometry concerns abstract objects and as such its status as knowledge cannot depend on how it is applied in physics. Euclidean geometry can retain other scientific applications, perhaps even as a theory of how we immediately experience the world through our sense organs and evolutionarily developed cognitive capacities (Hohol, 2019; Pantsar, 2021d). But its status as contextually a priori knowledge should be determined purely as a theory of mathematics.

Similarly, the Kuhnian notion of paradigm seen as contextually a priori knowledge goes into a divergent direction from my understanding. While arithmetic certainly gives us a possible way of experiencing the world, we can also make a much stronger claim. Given that proto-arithmetical abilities are universal, they give us an *inevitable* way of experiencing the world. Thus, arithmetic does not give us a paradigm that could be completely overthrown by a new paradigm. Of course, the paradigm can change, as it has done throughout the development of arithmetic, all the way from the first number concepts to modern axiomatic systems. But all such paradigm changes are constrained by proto-arithmetical abilities.

Arithmetic can never be overthrown by a system that conflicts with our basic cognition of numerosities based on proto-arithmetical capacities.

Understood in this way, contextual a priori knowledge refers to knowledge that is a priori in a *context set by empirical facts*. By this I do not mean that it is an empirical fact that we have the proto-arithmetical capacities that we do, even though this is obviously true. Nor do I mean that we can study proto-arithmetical capacities and the development of arithmetical cognition through empirical facts, even though this is obviously also true. What I mean is that when we learn basic rules concerning numerosities in our childhood, we do so by observing objects and organisms in our environment, that is, empirically. We could experience our environment in various different ways, but evolution has taken a trajectory which makes us observe discrete macro-level objects and their numerosities. When we first learn that two blocks and one block make three blocks – applying the object tracking system – we learn this similarly to the way we learn empirical facts. It is this developmental trajectory constrained by our proto-arithmetical abilities that determines the context for arithmetical knowledge. Once this context is set, however, we can limit empirical facts only to the context of discovery and remove them from the context of justification. That is, once the context is set, arithmetical knowledge is best understood as being a priori.

8.2 Objectivity

If arithmetical knowledge is contextually a priori in character, as argued in Section 8.1, the next important epistemological question concerns its *objectivity*. In Chapter 9, I will have more to say about the question of mathematical objects and their ontological status, but I see objectivity of mathematics primarily as an epistemological matter. In this, I follow a long line of tradition that Tait (2001) traces back to Cantor. As detailed by Cantor:

> [W]e may regard the whole numbers as real in so far as, on the basis of definitions, they occupy an entirely determinate place in our understanding, are well distinguished from all other parts of our thought and stand to them in determinate relationships, and thus modify the substance of our minds in a determinate way. (Cantor, 1883, §8)

It is clear that Cantor already had a clear distinction between ontology and epistemology in mind. For him, the reality of numbers concerns what place they take in our thinking, not their ontological status in some platonic sense. If the place numbers have in our thoughts is entirely determinate, numbers are real and knowledge about them objective. From a modern perspective, Cantor's account is, as pointed out by Tait (2001), perhaps too psychologistic. Instead of objectivity of mathematical thoughts, it is better to discuss objectivity of mathematical *discourse*. The important point for present purposes, however, is that objectivity is being discussed primarily in epistemological terms, and not ontological ones.[4]

The move to epistemology and the objectivity of discourses, however, does not mean that the matter becomes clearer. The first question to ask is when can we consider a particular discourse to be objective. Tait offers the following definition: 'Objectivity in mathematics is established when meaning has been specified for mathematical propositions, including existential propositions $\exists x F(x)$' (Tait, 2001, p. 22). But now the question is what is meant by 'meaning'. For Hilbert and the formalist approach of the early twentieth century, this was not a problem, since he thought that the meaning of mathematical propositions is fixed entirely by the axioms of the particular mathematical system (Zach, 2019). But after Gödel (1931) proved that all axiomatic systems of arithmetic are incomplete, this notion of meaning became untenable. However, I believe that it was untenable all along and Gödel's result simply provided one more problem with the formalist account. Even if axiomatic systems could be complete, we would still have the problem of the meaning of existential propositions, as pointed out by Tait. Axioms can merely tell us what kind of objects exist. They do not tell what it *means* to exist. For that, formalism needs to be connected to a philosophical account, such as conventionalism.

To begin the analysis here, I want to make clear that I accept the view that mathematical knowledge has a strong *appearance* of objectivity. Indeed, as argued in Pantsar (2021c), I believe that, together with mathematical applications in science, the appearance of objectivity is the main reason why mathematical knowledge is so often understood to be objective. The appearance of objectivity, however, should not be understood as any kind of *argument* for objectivity. This appearance can be compatible, for example, with strong conventionalist views that imply cultural subjectivity of mathematics.

As has become clear throughout this book, I do not believe that mathematical knowledge can be characterised in some way as subjective. But it is evident that the present epistemological account faces a challenge

[4] In the literature, this move from ontological to epistemological objectivity is sometimes referred to as 'Kreisel's Dictum', based on a remark by Dummett (1978, p. xxxvii) on Kreisel.

in avoiding subjectivity. Given my endorsement of arithmetic as the result of cumulative cultural evolution, we need to acknowledge the danger of losing the objectivity of arithmetical knowledge. When emphasising the culturally dependent aspects of arithmetic, at least *prima facie*, we face a potential tension with the widely shared view that arithmetical knowledge is objective.[5] It should be stressed that I am not claiming that arithmetical knowledge necessarily *is* objective in a strong sense. What I do believe is that arithmetical knowledge has a strong appearance of objectivity and this (together with the existence of arithmetical applications outside mathematics) is a *bona fide* philosophical problem that requires an answer.[6]

This challenge cannot be taken lightly, given that there are competing epistemological accounts that have straight-forward explanations of the apparent objectivity of arithmetic. Platonist philosophy of mathematics is based on the mind-independent existence of mathematical objects, which gives an immediate explanation for the apparent objectivity: arithmetical knowledge appears to be objective because it *is* objective, consisting of statements about mind-independent abstract objects. In Chapter 9 I will return to platonism, but it should already be disclosed that it would be extremely disappointing if the present account ended up requiring a platonist ontology or epistemology. In the spirit of Occam, I do not want to make unwarranted existential assumptions concerning mathematical entities. Consequently, if possible, I prefer an epistemological account that does not require a platonist ontology, while still providing an explanation for the apparent objectivity of arithmetical knowledge. Such an account exists, of course, in arithmetical *nativism*. If number concepts were innate and universally shared by humans, that would also explain the apparent objectivity of arithmetic. However, as I argued in Section 2.2, there is no empirical data to support such a nativist position. Indeed, the nativist accounts seem to be based on a confusion between proto-arithmetical and arithmetical abilities.

Given the problems of both platonist and nativist accounts, we cannot assume that arithmetical knowledge is objective in the strong sense that either mind-independent mathematical objects or innate number concepts determine it. However, as pointed out here and argued in detail in Pantsar

[5] When it comes to the question of just how widely this view is shared, unfortunately there is little strong data available. However, according to one online survey, 82.4 per cent of professional mathematicians believe in the objectivity of mathematical knowledge (Müller-Hill, 2009; for more on this topic, see Pantsar, 2015a).

[6] Here I will focus on the question of apparent objectivity. For the question of mathematical applications, see Pantsar (2018b, 2021c).

(2021c), the strong sense of objectivity is not the *relevant* sense for the present approach. What is needed is an explanation why arithmetical knowledge *appears* to be objective. As we have seen, in some cases, mathematical knowledge has this appearance because it consists of deeply entrenched conventions, such as the product of two negative numbers being positive. But, in the case of arithmetic, I submit, the appearance of objectivity comes primarily because arithmetical knowledge is based on evolutionarily developed proto-arithmetical abilities. Except for cases of developmental dysfunctions, these abilities are universal to human beings. This is what I meant earlier when I called arithmetical knowledge *maximally inter-subjective*. Proto-arithmetical abilities are not dependent on our languages or practices and, if the account developed in this book is correct, arithmetical knowledge and skills are in an important way constrained by proto-arithmetical abilities. It is, therefore, to be expected that arithmetic appears to be objective: in arithmetical cultures, the maximal inter-subjectivity of arithmetical knowledge makes it, for all practical purposes, objective.

It is of course possible that this objectivity is something stronger than just maximal inter-subjectivity. Penelope Maddy, for example, has argued that the foundations of arithmetic are a combination of set theory and logic, and that logic is based on our evolutionarily developed cognitive abilities. Thus, her account bears a clear resemblance to the one I have developed in this book. There are, however, important differences. Most importantly, in her view (based on proto-logic rather than proto-arithmetic) arithmetical knowledge is based on the logical structure of the world: 'Much as our primitive cognitive architecture, designed to detect [the logical structure of the world], produces our firm conviction in simple cases of rudimentary logic, our human language-learning device produces a comparably unwavering confidence in this potentially infinite pattern' (Maddy, 2014, p. 234). Given that I believe our primitive cognitive architecture to be the result of natural selection, I have a problem with the word 'designed' in this quotation. I see this kind of teleological terminology as misleading, but I think that Maddy's view can also be understood in a non-teleological way. Under a non-teleological reading, our primitive cognitive architecture *evolved* to detect the logical structure of the world. A similar argument has been presented by Helen De Cruz (2016), who argues that proto-arithmetical abilities have evolved to detect the structure of the world. Given that the structure of the world is ultimately responsible for the content of arithmetic, these accounts clearly provide an explanation for the apparent objectivity of arithmetical knowledge. But, unlike my account, their accounts make the realist assumption that the content of arithmetic somehow mirrors the structure of the world.

Here, however, I again want to remember the spirit of Occam and refrain from making unjustified and unnecessary assumptions. Putting a Kantian twist into the above views of Maddy and De Cruz, I do not see why proto-logic or proto-arithmetic would need to detect objective features of the world *an sich*. Instead, I support the weaker position that proto-mathematical abilities concern the structure that our primitive cognitive architecture imposes on our observations and thoughts. This may or may not mirror some objective structure of the world but, given that evolutionary advancements do not need to be based on accurate detection of the structure of the world, there is no need to make the realist assumption. I see this 'neo-Kantian' account as providing an equally strong explanation for arithmetical objectivity as the realist accounts of Maddy and De Cruz.

Indeed, it is important to distinguish my account from that of De Cruz (2016), who argues that only a realist ontology can explain how proto-arithmetical abilities have evolved. My counter-argument is that evolutionary endowments are simply vindicated by their adaptative success, which may or may not be connected to the objective structure of the world. If observing the world in terms of countable, discrete macro-level entities provides evolutionary advantages, we do not need to assume that the world is structured essentially in terms of such entities. Given that such evolutionary advantages are feasible in terms of acquiring food, avoiding predators, raising offspring, and many other crucial behaviours, we can have an evolutionary explanation based on proto-arithmetical abilities that makes no realist presuppositions like those of Maddy and De Cruz.

At this point, we should also examine the possibility that organisms could develop proto-arithmetical abilities that do not detect the structure of the world, but neither do the abilities detect some inevitable conditions that our cognitive architecture imposes on our observations and thoughts. In this case, proto-arithmetical abilities would simply be evolutionarily advantageous, perhaps for a reason that will remain unknown. While this is possible, it should be at least a plausible hypothesis that proto-arithmetical abilities are evolutionarily developed in tandem with observing the world in terms of discrete macro-level objects. This way, the proto-arithmetical abilities would be tied to the general way we experience our surroundings.

While I think that this is a feasible scenario, nothing in the present epistemological account ultimately depends on it. What I have wanted to show is this: the argument for objectivity of arithmetical knowledge based on our primitive cognitive architecture is similar regardless of whether or not we connect it to a realist account. Proto-arithmetical abilities could be 'truth-tracking' in that they ensure that our cognitive capacities provide us with accurate information about the structure of the world. But they could also be evolutionarily advantageous for some other reason. In either scenario, we have an explanation for the apparent objectivity of arithmetical knowledge, based on it being maximally inter-subjective. And, as I have argued, the resulting notion of objectivity is strong enough to account for the widely shared impression that arithmetical truths are objective.[7]

Not everyone agrees with this line of thinking. Daniel Hutto, for example, criticises the constructivist account of Lakoff and Núñez (2000) for not being able to offer a strong enough account of objectivity:

> Lakoff and Núñez (2000) take [the basis of mathematics on embodiment] to be a virtue of their account; they are satisfied in rejecting the romantic ideal [of realism] just so long as they can avoid endorsing the idea that the meaning of mathematics is generated by arbitrary social conventions. However, their account of the truth conditions of mathematics only secures that it is non-arbitrary with respect to our shared embodiment – hence it falls a long way short of being an account of the objectivity of mathematics. (Hutto, 2019, p. 835)

Thus, while escaping the arbitrariness of strict conventionalist accounts, Hutto claims that the account of Lakoff and Núñez fails to respond to another accusation of arbitrariness, namely that of it being an arbitrary matter that humans happen to share a particular type of embodiment. To avoid this type of arbitrariness of mathematical truth, Hutto (2019) argues that we need to combine the conceptual constructivist account with a realist account of mathematical truth.

This argument, however, is quite problematic. If we invoke a platonist (or other realist, such as physicalist) account of mathematical truth, we face all the problems that the present epistemological account has been designed to avoid. Fortunately, the problem Hutto points out does not seem to be quite as serious as he suggests. After all, we are not trying to explain the objectivity of arithmetical knowledge in a strong, realist sense. The epistemological view I am proposing does not claim that arithmetical knowledge is objective in a mind-independent manner. Instead, it claims that arithmetical knowledge is objective in a manner that is not strictly

[7] For a more detailed argument, including an analysis of the notion of 'objectivity' in terms of Wright's (1992) definition, see Pantsar (2021c).

conventional. Therefore, the threat of arbitrariness with regard to our embodiment is not a real threat. The trajectories of development of arithmetic in different cultures have shown that at least basic arithmetic is not merely a matter of conventions, and the shared proto-arithmetical origins of arithmetical systems can explain that. Thus, while it may be an arbitrary matter that there are organisms with proto-arithmetical abilities sharing a particular type of embodiment, in the context in which such organisms exist, the content of arithmetic is *not* arbitrary. Instead, it is constrained by evolutionarily developed capacities. This, I contend, shows that arithmetical knowledge – characterised as contextually a priori – is objective in a strong enough sense.[8]

8.3 Necessity

After apriority and objectivity, the next characteristic traditionally associated with arithmetical and other mathematical knowledge is necessity. This, of course, is closely connected to the other two characteristics treated so far. It seems feasible that, if arithmetical knowledge were both objective and a priori, it could not be contingent. In this line of thinking, objectivity entails that the status of arithmetical truths is mind-independent and apriority entails that these truths can be established independently of sensory experience. It is difficult to see how such objective truths concerning abstract objects or structures could be contingent. Indeed, as Kripke (1980, p. 35) pointed out, in the time before his *Naming and Necessity*, few philosophers even made the Kantian distinction between a priori and necessary.

While the view that mathematical truths are necessary has been very popular, spelling out *why* this should be the case has proven to be difficult. In the traditional logicism of Frege, Whitehead and Russell (Frege, 1884; Whitehead & Russell, 1910), this question would have been easier, given that all mathematical truths were understood to be logical truths. But, given that we need at least some non-logical axioms in mathematics, such as Hume's Principle in the case of arithmetic (see, e.g., Hale & Wright, 2001; Linnebo, 2018b), mathematical truth cannot be completely derived from logical truth (not everyone agrees; see, e.g., Hintikka, 1996).[9]

[8] The discussion on Hutto is connected to the question of whether arithmetical cognition can be explained by enactivist, in particular radical enactivist, accounts. See Pantsar (2023c) for a detailed discussion on that topic.

[9] See also Yablo (2002) for an argument that arithmetical truths are in fact (first-order) logical truths.

For this reason, in modern approaches to mathematical necessity, other grounds are needed (for a set-theoretic account of the metaphysical necessity of pure mathematics, see Leitgeb, 2020).

Here it is not possible to give a detailed treatment to the general question of mathematical necessity, but fortunately only a brief exposition is needed in order to have sufficient theoretical tools for treating necessity in the context of the present account. Ever since the second half of the twentieth century, it has been commonplace to treat necessity in terms of modal logic and possible world semantics (see, e.g., Kripke, 1963; Lewis, 1970). Under this approach, necessary truths are true in all possible worlds. Standardly, arithmetical truths have been used as canonical examples of such necessary truths: while we can envision all kinds of possible worlds, the argument goes, we cannot envision one where it is not the case that $2 + 2 = 4$. For the present epistemological account, however, this is potentially problematic. Given that the epistemological account I have developed is based on proto-arithmetical abilities, any possible world where there are no organisms with proto-arithmetical (or arithmetical) abilities would seem to be such a world. If the truth of $2 + 2 = 4$ depends on the existence of agents with proto-arithmetical (or arithmetical) abilities, we face the troubling prospect that in some possible worlds it is not a truth.

One possible solution to this problem would be that, while arithmetical knowledge arises from proto-arithmetical abilities, arithmetical *propositions* could have truth values independently of whether there are agents that have knowledge of them. This solution, however, requires some kind of realism over mathematical truth-values. Arithmetical truths would be mind-independent, which puts the status of the contextual a priori account into question. While this is by no means in conflict with the epistemological account I have developed in this book, it certainly goes against the Occamian spirit of ontological and epistemological parsimony that is central to my project.[10]

[10] This is not to suggest that arithmetical propositions whose truth values are unknown *lack* truth values, which was something one reviewer of this book was worried about. I do believe that unproven propositions like the Goldbach conjecture have determinate truth values. I see no reason to believe that the truth value of the Goldbach conjecture, for example, is mind-independent, given that it would not exist as an arithmetical proposition without the human practice of arithmetic. However, this does not imply in any way that in our arithmetical systems, like the Dedekind–Peano axiomatisation, Goldbach conjecture does not have determinate truth value even if we do not know it yet.

Another possible solution to the problem would be to argue that, while $2 + 2 = 4$ may not be true in all possible worlds, this does not imply that $2 + 2 = 4$ could be *false*. According to this argument, it could have been the case that proto-arithmetical abilities never emerged. But if they *did*, then $2 + 2 = 4$ would be necessarily true. In this scenario, either arithmetical truths are the way they are, or there are no arithmetical truths.[11] While this argument is appealing, it also makes an ontological claim that is too strong for the spirit of my project. In an account like the present, in which arithmetical knowledge is (partly) based on our basic cognitive architecture, we cannot rule out the possibility that this architecture could have evolved in a radically different manner. In this case, we could have some kind of number concepts and arithmetic, but they could be different. Any argument to the effect that arithmetical knowledge must develop among the lines it has in our actual world suggests a more robust ontological status for arithmetic: namely, that arithmetical truths somehow determine the way biological organisms must develop. This is a stronger ontological claim than I am prepared to make.

A weaker form of this argument is the one provided by Maddy and De Cruz, who argue that our cognitive architecture has developed to detect the structure of the world (De Cruz, 2016; Maddy, 2014). However, as detailed in Section 8.2, this also includes an assumption that goes against the spirit of my project. This is the assumption that our cognitive architecture has developed to detect objective features of the world rather than to simply enhance survival and reproduction. While it is very much possible that the two are closely connected, this is not something we can simply assume.

With these problems, is there a way to establish the necessity of arithmetical truths in the present account? As with apriority and objectivity, I propose that there is, but we need to understand necessity in a proper context. I argue that this context can be understood in terms of the seminal work of Kripke (1980). Kripke's main achievement there was to divorce the notions of a priori and (metaphysical) necessity from each other, most importantly showing how also a posteriori knowledge can be necessary. The best-known example of this is the sentence 'Hesperus is

[11] This distinction can be framed in terms of epistemic and metaphysical necessity. While metaphysical necessity concerns objective states of affairs, epistemic necessity concerns our knowledge. In Kant scholarship, there is debate over what kind of necessity he saw for arithmetical statements (see, e.g., Hebbeler, 2015). For my purposes, the Kantian connection notwithstanding, both forms of necessity are too strong for arithmetic.

Phosphorus'. Both 'Hesperus' and 'Phosphorus' are proper names and as such what Kripke called *rigid designators*, which pick out the same thing in every possible world where that thing exists. For 'Hesperus', that thing is the evening star. For 'Phosphorus', it is the morning star. However, both proper names pick out the same thing, namely the planet Venus. Hence, the sentence 'Hesperus is Phosphorus' is necessarily true, because in all possible worlds 'Hesperus' and 'Phosphorus' refer to the same thing. Therefore, given that the fact that both names refer to the planet Venus was discovered empirically, Kripke argues, 'Hesperus is Phosphorus' is a case of necessary a posteriori knowledge.

Could we use this kind of argument to save the necessity of arithmetical knowledge in the contextual a priori account I have developed in this book? After all, while contextual a priori knowledge in my account is certainly not a posteriori, it is not strictly a priori either. Given that Kripke extends the notion of necessity beyond the realm of strictly a priori, this strategy seems promising. However, there is one big difficulty. Kripke introduced the term 'rigid designator' for terms that pick the same object in all possible worlds in which that object exists. But how should we understand this semantic idea of rigidity, that is, how are proper names connected to their objects? One suggested way has been to evoke a causal theory of reference. We can know that the term 'Hesperus' picks out the evening star in every possible world where the evening star exists because there is a causal link from the name 'Hesperus' to the event of giving it its reference. In the case of arithmetical knowledge, however, we cannot assume that there are such causal links to already existing things, that is, natural numbers. Doing so would again go against the spirit of my project, which aims to provide a feasible epistemological theory without assuming the existence of mind-independent numbers. Besides, as abstract objects, natural numbers are supposed to be causally inactive, which would make causal links to them in principle impossible.

However, as established in Section 8.2, even if we do not assume the existence of mind-independent objects, we can still retain a strong sense of *objectivity* for arithmetical knowledge. Natural numbers may not be objects in the same sense as the planet Venus is, but there is definitely a strong sense of objectivity in the way we use numeral symbols and numeral words. The reason is, as I have argued in this book, that we share number *concepts* based on universal proto-arithmetical abilities. This, I submit, is a strong enough sense of objectivity for a Kripke-type theory of reference. In the present account, the statement $2 + 1 = 3$ is true in all possible worlds where enough arithmetical or proto-arithmetical ability has developed to give reference for the numerals '2', '1' and '3'.[12]

Do numerals designate rigidly enough to save necessity in the contextual a priori account of arithmetical knowledge? It might seem that if we do not include mind-independent arithmetical objects in our ontology, numerals cannot be rigid designators. But I believe that this is mistaken. In the present account, numerals are thought to refer to something strongly objective, that is, the referents of shared number *concepts* developed based on our proto-arithmetical abilities. Since these abilities are constrained by our biological structure, this objectivity is very strong indeed. While there is no single mind-independent object that the numeral '2' designates, it still ultimately manages to pick out something that is at the basis of animal and infant behaviour, as well as of our arithmetical theories. Therefore, the numeral '2' is connected to that 'something' in every possible world in which biological organisms develop along the lines of biological evolution that took place in the actual world.

What that 'something' is in the case of arithmetic is more complicated than in the case of Venus. In this book, I have argued that the thing that numerals pick out are the referents of culturally shared number concepts. Indeed, this is *all* that numbers as arithmetical objects are assumed to be in this present account. While this may not make numerals rigid designators in the sense that they pick out a unique object in all possible worlds, I submit that it comes close enough for the present purposes. As in the case of objectivity, our real target phenomenon is not a platonist, metaphysical notion of mathematical necessity. Instead, we need to explain the *apparent* necessity of arithmetical truths. I believe that this can be done.

There are, of course, possible worlds in which arithmetical knowledge does not develop. But, in Kripke's framework, it is enough that the name picks out the same thing in all possible worlds where that thing exists. What I have tried to show here is that, when adjusted to our understanding of arithmetical objects as referents of culturally shared concepts, a Kripke-type theory is compatible with the present account of epistemology of arithmetic. For that, it is enough that a numeral picks out the same thing in all possible worlds where sufficiently similar biological organisms develop, when compared to the actual world. Given that this domain of possible worlds includes any world in which humans as biological subjects exist, in this modified Kripkean framework we can see why arithmetical

[12] Here I use numeral symbols but, as we have seen, standardly numeral words come first in both ontogeny and cultural history.

truths appear to be necessary. The only problem is that, in the case of numerals, the 'things' that they refer to – that is, the referents of culturally shared number concepts – are not unique, mind-independent objects. However, this does not mean that the reference of numerals is indeterminate, given what we learned in Part II about culturally shared number concepts. Thus, I contend that numerals are rigid enough designators to account for the apparent necessity of arithmetical statements.[13]

To end this section, we should give attention to the fundamental meta-semantic question concerning the above account: how do numerals refer? Is the reference causal or of some other type? As pointed out by Stalnaker (2003), we should distinguish the Kripkean notion of rigid designation from the causal theory of reference, even if the two are often connected. Here I have argued that numerals according to the present account can be 'rigid enough' designators. But how do they refer to their objects, that is, the referents of culturally shared number concepts? Generally, I think that a causal theory of reference is compatible with my account. Naturally, the process is much more intricate than the basic 'dubbing' in which a word is associated with its reference (for example, naming a kitten). But I believe that in principle – and only in principle – it is possible to trace a causal chain in how the word 'three' came to have the reference of the culturally shared number concept. This history involves the establishment of cultural linguistic and arithmetical practices based on the proto-arithmetical abilities, and it cannot be traced in detail. However, in Part II of this book I hope to have provided a plausible framework for such histories.[14]

8.4 Universality

The final characteristic traditionally associated with mathematical knowledge is its universality. Arithmetical and other mathematical knowledge is thought to apply everywhere and in all circumstances. Again, the present contextual a priori account may at first glance appear to conflict with that. After all, if arithmetical truths are universal, how can we account for cultures that do not possess arithmetical skills or knowledge, let alone the possible worlds in which not even proto-arithmetical abilities evolved for organisms? A quite understandable initial intuition would be that even

[13] For more details concerning this argument, see Pantsar (2016b).

[14] Even though I believe my account to be compatible with a causal theory of reference, I do not see it as committing to a particular meta-semantic theory. Unfortunately, it is not possible to give this topic a detailed treatment here. For more on meta-semantic theories, see Michaelson and Reimer (2022).

in those worlds $2 + 2 = 4$ is true and $2 + 2 = 5$ is false. However, as we have seen in Section 8.3, this intuition, the *appearance* of necessity, can be explained satisfactorily in terms of arithmetical knowledge developing necessarily among similar lines, at least when basic arithmetical operations with finite numbers are considered. In this section, I submit that the appearance of universality can be explained in a similar manner. But, in order to do this, it is important to first recognise that the explanandum is the *apparent* universality, not any realist notion of mathematical universality.

Can the contextual a priori account of arithmetical knowledge explain the apparent universality of arithmetical knowledge? It may seem like this is ultimately the same question as the one treated in Section 8.2, namely the question of apparent *objectivity*, or the one treated in Section 8.3, namely the question of apparent *necessity*. However, there are important differences between the three questions. The question of necessity was analysed in terms of possible worlds, but the question of universality concerns universality in the *actual* world. The argument I presented was based on arithmetical knowledge developing in converging manners in every possible world where organisms with proto-arithmetical (or arithmetical) abilities develop. But while this explains, as I hope to have shown, the apparent necessity, it leaves open the possibility that in the actual world arithmetic does not apply universally. It could be argued, for example, that arithmetical knowledge then applies only in parts of the world where there are organisms with proto-arithmetical or arithmetical abilities.

The question of apparent objectivity would appear to be more closely intertwined with the question of apparent universality. Yet also those two questions need to be distinguished. Apparent objectivity still leaves room for appearance of non-universality. This is the case certainly with arithmetical practices, such as methodology, cognitive tools and numeral systems, but also with the content of arithmetic. We have seen that infinity, for example, has been treated fundamentally differently in different arithmetical cultures (Ifrah, 1998; Merzbach & Boyer, 2011). The status of zero as a natural number varies even in contemporary mathematics. There is no universal answer among mathematicians to the question 'is zero a natural number?' and, consequently, neither is there *apparent* universality. One could still make the case, however, that the question of whether zero is a natural number has an *objective* answer: either it is a natural number or it is not. In this line of thinking, there is a definite answer to arithmetical questions involving zero, we just need to find that answer. While it is important to recognise that the questions of apparent objectivity, necessity and universality are different, clearly, they are also closely connected. Indeed, as I am going to argue next, with proper amendments, the treatment of apparent objectivity can also be applied to the question of apparent universality.

The key to making these amendments is to recognise the role of conventions in arithmetic. The disagreements that make systems of arithmetic differ, I submit, result from different conventions. Some of these conventions are canonised, as in the ISO 80000 standards of quantities and units, which state that zero is the first natural number.[15] But, nevertheless, they are still conventions, providing answers by decree rather than through some objective mathematical explanation. For this reason, I submit that we need to focus on the part of arithmetic that is *not* based on conventions, but rather on our proto-arithmetical abilities. While there are different answers to questions like 'is five minus five a natural number?' there are no different answers to questions like 'what is five plus five?' The latter question, belonging to basic operations on finite (non-zero) natural numbers always gets the same answer – if the question makes any sense at all. In cultures like Pirahã and Munduruku it does not make sense, but that is because they are non-arithmetical cultures, not because they have a different system of arithmetic (Gordon, 2004; Pica et al., 2004). There is no reason to believe that any culture, if they develop arithmetic based on proto-arithmetical abilities, will end up with divergent arithmetic of basic operations and finite numbers.

Therefore, it is important to identify what the relevant arithmetical content is. While there have been divergent arithmetical systems with regard to the status of zero and infinity, there have been none that diverge with regard to basic operations and finite numbers. In this book, I have tried to provide an explanation why this is the case, based on our proto-arithmetical abilities determining the ontogenetic, phylogenetic and historical trajectory of arithmetical development. This determination is not total, as we have seen. But it is, as I argued in Section 8.3, strong enough to account for the apparent objectivity of arithmetical knowledge. This argument can be directly applied also to the apparent *universality* of arithmetical knowledge, when we remember the data strongly supporting the view that the proto-arithmetical abilities are universal in humans (Dehaene, 2011; Feigenson et al., 2004; Pica et al., 2004; Spelke, 2011). Arithmetical knowledge appears universal to us because humans

[15] www.iso.org/standard/30669.html

universally share their proto-arithmetical abilities and arithmetic is developed on those abilities.[16]

This does not mean, of course, that arithmetical knowledge is universally present. There are non-arithmetical cultures that do not share our knowledge even for small numerosities. But it could also be that other lifeforms (including perhaps alien ones) do not share anything like our proto-arithmetical abilities. Their basic notions of numerosity could be, for example, continuous ones. As seen in this chapter, some philosophers have argued that the structure of the world is logical or arithmetical, which explains why the proto-logical or proto-arithmetical abilities have evolved (De Cruz, 2016; Maddy, 2014). I have already stressed that I am not satisfied with this argument. All we can confidently say about the origin of proto-arithmetical abilities is that they have been evolutionary advantageous. Otherwise, it would be puzzling why they have evolved for so many animals. But this advantage could be due to other reasons, and not because the abilities somehow detect the structure of the world. In another line of evolution, it is at least conceivable that organisms could develop different abilities for numerosities that prove evolutionarily advantageous. In the presence of such organisms, we could not maintain that the proto-arithmetical treatment of quantities is universal.

To make sense of such scenarios, it is important to distinguish the universal *presence* of abilities from the universal *applicability* of abilities. Arithmetical abilities are not universally present, and it could be the case that organisms evolve that have abilities for treating numerosities that are fundamentally different from proto-arithmetical abilities. But neither of these observations makes a difference to the question of why arithmetical knowledge appears to be universal to us. We have seen that arithmetic appears universal to us for an important part because humans share proto-arithmetical abilities. To conclude this chapter, I want to focus on the question of applicability. Aside from the universality of proto-arithmetical abilities, an important reason for the appearance of universality of arithmetic is that there does not seem to be any temporal or spatial limits to the applicability of arithmetical knowledge. Whatever we can do with arithmetic, it appears that we can do it anywhere.

Contrary to De Cruz (2016), I do not want to pursue the explanation that the applicability is due to the universe being arithmetically structured. Rather, I want to make the Kantian move to argue that it is due to how our

[16] By 'us', I primarily refer to the Western/Westernised world and the way we understand arithmetic. However, there is no reason to think that this appearance is limited to Western cultures.

observations of the universe are structured. As long as we observe macro-level discrete objects, our proto-arithmetical abilities will be applied to them. It could be that there are parts of the universe, for example gas planets, where such observations would be rare. But, nevertheless, *if* we would observe discrete macro-level objects in those environments, our proto-arithmetical abilities would be applied. For people in arithmetical cultures, it would also be possible to apply their arithmetical knowledge. If this thought experiment is seen as plausible, as I believe it should be, it should be no wonder that arithmetical knowledge appears to be universal to us. In addition, under an understanding in the appropriate context, arithmetical knowledge would appear to be a priori, objective and necessary. Thus, I have argued in this chapter, arithmetical knowledge would retain all the characteristics traditionally associated with it. This is an important strength of the present account.

Finally, we are in the position to also address the question that was important for Kant, namely, whether arithmetical knowledge is synthetic or analytic. I have defended the view that arithmetical knowledge is not strictly a priori, but it is contextually so. It is this context-dependency that is key to answering the question about syntheticity and analyticity. In the context of arithmetically skilled people, it may appear that arithmetical knowledge is analytical. After all, it follows from a small group of axioms whose meaning determines the truth of all arithmetical propositions.[17] But, again, this would be missing the bigger picture. The axioms themselves are not arbitrary conventions. Instead, we have established that arithmetical truths are (partly) constrained by our proto-arithmetical abilities. In arithmetical cultures, people started to accept arithmetical truths because they conformed to our proto-arithmetical abilities. While our modern formal systems of arithmetic have obviously come a long way from these early stages of arithmetic, they are still part of the same development. The early stages in the history of arithmetic were closely connected to experiencing the world in terms of our proto-arithmetical abilities. The early stages of ontogeny of arithmetic still are. Hence, arithmetical knowledge is not purely analytic: it is not only about the meaning of the axioms and theorems derived from them. It is also about the way we observe the world through our evolutionarily developed cognitive architecture, which includes proto-arithmetical abilities. In this sense, arithmetical knowledge can be understood as synthetic. Certainly,

[17] With Gödelian restrictions on the completeness of formal axiomatic systems of arithmetic, see Gödel (1931) and Pantsar (2009).

this sense on syntheticity is much weaker than Kant had in mind. But it is enough to challenge the notion that arithmetical knowledge is purely analytic.

8.5 Summary

In this chapter, I dealt with the threat that the present account strips arithmetical knowledge of all the important characteristics traditionally associated with it: apriority, objectivity, necessity and universality. I argued that apriority can be saved in the strong sense of arithmetical knowledge being *contextually* a priori in the context set by our cognitive and physical capacities. Objectivity can be saved in the sense of maximal intersubjectivity, while necessity can be saved in the sense of arithmetical theorems being true in all possible worlds where cognitive agents with proto-arithmetical abilities have developed. Finally, universality of arithmetical truths is saved through arithmetic being universally applicable and shared by all members of cultures that develop arithmetic based on proto-arithmetical abilities.

CHAPTER 9

Ontological Considerations

After the epistemological considerations of Chapter 8, we are finally ready to turn to perhaps the historically most discussed topic in the philosophy of mathematics: what are natural numbers as mathematical objects *like*? Thousands of pages have been written on that issue, so it is not possible here to give the topic a comprehensive treatment. In this chapter, I am not concerned, for example, with the question of whether numbers exist ultimately as sets and, if so, how numbers should be defined as sets (see, e.g., Benacerraf, 1965). The reason for my choice is simple. As I have tried to show in this book, and will continue to do in this chapter, I believe that we can have a satisfactory epistemology of arithmetic without assuming the existence of mind-independent mathematical objects. And, if that assumption is not needed, it makes little sense to argue whether sets or numbers are ontologically primary mathematical objects. If neither are mind-independent, then their ontological status is always secondary to human cognitive and cultural practices.

However, even if my project were successful, namely if I could show that we do not need to assume the mind-independent existence of natural numbers, it would not follow that natural numbers do *not* exist in a mind-independent fashion. In the most general way, this concerns nothing more than the impossibility of disproving the existence of something. Even if my epistemological and ontological considerations were completely satisfactory, there would still remain the possibility that natural numbers exist also in *another* way, which is perhaps not connected at all to the way humans get knowledge of them. That kind of realist position cannot be disproved and, hence, I do not find it particularly interesting. However, here I recognise that realism concerning mathematical objects is not simply a case of this general challenge, akin to disproving the existence of Santa Claus or the Loch Ness Monster. There are some genuinely persuasive reasons for believing in the mind-independent existence of mathematical objects. Ultimately, as I will try to show in this chapter, none of those reasons are needed in the present account. But they do warrant consideration. Indeed, I believe that, in weaker forms of realism, such as those of 'thin objects' (Linnebo, 2018b) and 'subtle platonism' (Rayo, 2013), the present account can help explain why they seem so appealing.

In this chapter, I will build on the notion I have already discussed, namely that natural numbers as objects exist in a specific mind-dependent way, namely as referents of *culturally shared number concepts*. They may exist in other ways, as well, but I argue that in the present account no other type of existence needs to be assumed. Thus, the ontology of mathematics I advocate is minimal. However, this does not mean that I see the ontology of mathematics as an uninteresting topic. Indeed, I hope that the present account comes with many fruitful consequences for the ontological study of mathematics, rather than as a rejection of all things ontological in the philosophy of mathematics.

9.1 Realism

Realism over mathematical objects is standardly understood as platonism. As explained in the Introduction, here Platonism with a capital 'P' refers specifically to Plato's philosophy, while platonism with a lower case 'p' refers to the general realist metaphysical position on mathematics, according to which mathematical objects exist mind-independently (Balaguer, 2016). There have been suggestions that it would be better if we replaced the term 'platonism' with 'realism' (e.g., Tait, 2001). While I partly agree with that, especially due to the distance of some realist positions from Plato's philosophy, that suggestion comes with a problem. While the standard interpretation of platonism is that mathematical objects are *abstract*, that is, non-spatial, non-temporal and causally inactive, there can also be a realist ontology that takes the objects of mathematics to be ultimately *physical*. Due to the possibility of such *physicalist* theories, it is better to use the term 'platonism' in the traditional sense, while using the term 'realism' to include both platonism and physicalism.

Physicalism of mathematics is a sub-field of the general metaphysical physicalist view that everything is physical (see, e.g., Stoljar, 2022). To oversimplify the matter, but only slightly, the mathematical physicalist view is that, if there were something non-physical it would include mathematical objects. But, since we can also explain mathematics in a physicalist framework, the argument tends to go that we can advocate general physicalist metaphysics. Thus, I understand physicalism in fact as one type of *nominalism* in the ontology of mathematics, in that it denies the existence of *abstract* mathematical objects. Nominalism comes in many forms but, to the best of my knowledge, nobody advocates the view that mathematical objects are a specific type of physical objects. There are accounts according to which numbers are properties of relations of physical objects (see, e.g., Bigelow, 1988), but this should be distinguished from thinking of mathematical objects as being physical objects. Thus, the typical mathematical physicalist view is that mathematical *truths* are ultimately grounded in the physical world. We have seen that Maddy (2014) and De Cruz (2016), among others, have supported such a view.[1]

I have already discussed those views in detail, and here I will not focus more on physicalism over mathematics. As I have argued before, I do not see a reason why our mathematical theories need to be assumed to detect the structure of the physical world. Even though many types of physicalist views have been discussed in the literature (for a good, although by now dated selection, see Irvine, 1989), ultimately they come down to the same problem. How can we know that the structure of our observations corresponds to the structure of the world? As I have argued in this book, this is an assumption we do not need to make to account for arithmetical knowledge. It is enough that the structure of observations is evolutionarily advantageous, which may or may not be due to them detecting the structure of the physical world. Therefore, a physicalist realist interpretation adds nothing to the present epistemological account that cannot be explained without making the realist assumption. For that reason, I will focus here on platonist realism.

Platonist views can be divided into two types of positions. According to the first one, associated with Plato's original philosophy, mathematical objects such as numbers exist in a mind-independent world of ideas (Plato, 1992). According to the second one, mathematical objects exist as places in mind-independent mathematical *structures*. In the structuralist view, independent objects are not considered to be the subject matter of mathematics. Instead, mathematics concerns structures, like those of natural numbers or real numbers. This way, the number two, for example, is the second position in any instantiation of the natural number structure, rather than any structure-independent object.

[1] It should be specified, however, that, while both of them refer to their position as realism, neither explicitly advocates physicalism. In this, the account presented by Bigelow (1988) is different, since it is explicitly physicalist. As I understand the relevant notions, nominalism and physicalist realism are compatible notions when it comes to philosophy of mathematics. This is obviously different with regard to nominalism and platonism.

The platonist version of structuralism is called *ante rem* structuralism in the literature (Shapiro, 1997). The idea behind *ante rem* structuralism is that mathematical structures, like the natural number structure, exist regardless of whether there are systems of objects structured in that way. This contrasts it with *modal* structuralism, according to which mathematical structures are modal entities, in that they exist in the sense that it is possible for a structure of objects to instantiate them (Hellman, 1989; Shapiro, 1997, p. 10). A related view is called *in re* structuralism in the literature. According to this view, abstract structures exist in their concrete instantiations.[2] According to *in re* structuralism, if we destroy all objects structured like natural numbers, the natural number structure itself ceases to exist. But, given that for many mathematical structures (although not necessarily natural numbers), there may not be enough physical objects to instantiate them, *in re* structuralism seems to necessitate the existence of abstract objects, and is thus similar to Platonism (i.e., Plato's ontological account) (Shapiro, 1997, pp. 9–10).[3] Hence, we can treat Platonism and *ante rem* structuralism as the two main traditional platonist views in the philosophy of mathematics.

Are Platonism and *ante rem* structuralism compatible with the epistemological approach of this book? Clearly the answer is yes in both cases. I have argued that arithmetical knowledge concerns numbers that exist as referents of culturally shared concepts. It is perfectly possible that these number concepts have their ideal counterpart in a Platonic mind-independent world of abstract objects. Similarly, the number concepts can have platonic counterparts in the positions of an *ante rem* structure of natural numbers. The important question, however, is whether making such ontological postulations causes any increase in explanatory power. In this book, I have argued that all knowledge of natural numbers can be based on proto-arithmetical abilities and the way they are culturally developed into arithmetical theories. But, so far, we have not tackled in detail the question of whether there can be some *other* way of gaining knowledge of natural numbers. If there is not, then it is difficult to see the motivation for platonist ontology, whether traditional Platonism or *ante rem* structuralism.

[2] It is a matter of debate how different modal and *in re* structuralism are (for more, see Reck & Schiemer, 2023).

[3] The fourth interpretation of structuralism, according to Shapiro (1996), is *post rem* (or eliminative) structuralism, according to which mathematical structures have no existence and the talk of structures refers only to common properties between systems. Eliminative structuralism is, thus, an essentially nominalist view.

As mentioned in Chapter 8, there have been well-known proponents of such an alternative epistemic access to mathematical knowledge as part of a platonist ontology. Gödel (1983) believed in a special epistemic faculty for mathematical knowledge. More recently, Penrose (1989, 1994) has argued that a special form of mathematical 'seeing' can give us mathematical knowledge. However, to the best of my knowledge, there is no empirical support for such special epistemic access. In the line of research on the psychology of mathematical invention and thought following Hadamard (1954), there is nothing to suggest that mathematical intuitions form an epistemic capacity that cannot be explained in terms of other cognitive capacities. Granted, this topic is notoriously difficult to study empirically, but brain-imaging data of professional mathematicians certainly do not support a separate cognitive capacity. Instead, as mentioned in Section 3.4, mathematicians have been detected to activate a brain network consisting of bilateral frontal, intraparietal and ventrolateral temporal regions when processing mathematical statements (Amalric & Dehaene, 2016). This network overlaps with the network consisting of the bilateral prefrontal cortex, intraparietal sulcus and inferior temporal cortex, which is the network associated with proto-arithmetical abilities (Nieder, 2019). This suggests the possibility that the mathematical intuitions described as 'seeing' or some similar term are closely connected to the proto-mathematical origins. While more research is needed, the current state of the art in empirical research does not support any epistemic access to mathematical objects that cannot be explained in the present framework.

Not everybody agrees, however, that the proto-mathematical roots of mathematics should be seen as important for the development of mathematical knowledge. Recently, a realist ontology of mathematics has received surprising support from the radical enactivist Daniel Hutto, who has accused accounts like that of Lakoff and Núñez (2000) of 'neural fetishism' (Hutto, 2019, p. 830). In outlining an enactivist account of mathematical cognition, Hutto writes:

> Subtract any residual commitment to mental representation, information-processing stories, and neuro-fetishism. Add, in place of these items, a more Andersonian account of neural reuse – one that focuses on the pluripotent, protean brains and which places the greater weight on the contributions of socio-cultural practices in establishing mathematical content and competencies ... Subtract any residual constructivism, anti-realism, and idealistic elements from the account. Finally, subtract any lingering psychologism about mathematics and its content. (Hutto, 2019, p. 835)

This recipe seems like a bad fit with the present approach. Not only have I been advocating a form of psychologism in relation to proto-arithmetical abilities, but my account is fundamentally anti-realist, given that no realist assumption about mathematical objects or structures is required. Indeed, interpreted in an anti-realist manner, Hutto's account seems to be in danger of becoming conventionalist. If all psychologism is subtracted, what is there to prevent mathematics being about arbitrary conventions? However, Hutto sees things differently and argues that it is *psychologism* that cannot avoid arbitrariness in a satisfactory manner:

> Lakoff and Núñez (2000) ... are satisfied in rejecting the romantic ideal just so long as they can avoid endorsing the idea that the meaning of mathematics is generated by arbitrary social conventions. However, their account of the truth conditions of mathematics only secures that it is non-arbitrary with respect to our shared embodiment – hence it falls a long way short of being an account of the objectivity of mathematics. (Hutto, 2019, p. 835)

Hence, Hutto ends up endorsing a realist position over mathematical truth, arguing for the possibility that the subject matter of mathematics is objective and mind-independent (Hutto, 2019).

However, I am not convinced that an approach like the present one requires any realist assumptions to account for the objectivity of mathematics. As argued in Section 8.2, I contend that our shared embodiment and the shared proto-arithmetical systems are enough to save objectivity of arithmetic in a strong enough sense. Indeed, if we make the move that Hutto suggests and endorse realist mathematical truth, I believe we would be abandoning the Occamian spirit of this book. Whether the realist position is understood in platonist or physicalist terms, we would be making an assumption that has no basis in our best empirical understanding of the development of arithmetical cognition. Thus, the trade that Hutto suggests, that is, removing psychologism and introducing realism, would be hard to motivate. And if objectivity of arithmetical truth is the main reason for that trade, based on the consideration in Section 8.2, we need not worry. Objectivity can also be saved in a strong enough sense without making any realist assumptions.

9.2 Thin Objects and Subtle Platonism

In the present approach, the main reason for not subscribing to platonist ontology is that there is no reason to do so. It is impossible to establish that mind-independent mathematical objects do not exist, but fortunately we can have an adequate epistemology of arithmetic without them. In the Occamian spirit of minimising ontological commitments, we can then conclude that a frugal ontology is preferable to a platonist ontology. However, in the philosophical literature, there have also been platonist accounts that are presented in the spirit of minimising ontological commitments. Here I will focus on two such accounts, Linnebo's (2018b) 'thin objects' and Rayo's (2013, 2015) 'subtle platonism'.

Linnebo has defined thin objects as such that their existence 'does not make a substantial demand of the world' (Linnebo, 2018b, p. 4). More specifically, thin objects exist as referents of singular terms in abstraction principles. As explained in Section I.1.3, Frege (1884) noted that analytic a priori truths can make existential claims. In his example, directions can be introduced as new abstract objects by following analytic truth:

> The direction of line A is the same as the direction of line B if and only if A and B are parallel.

In a similar manner, the existence of natural numbers as abstract objects follows from Hume's Principle (together with logic), which states that the number of *F*s is equal to the number of *G*s if and only if the *F*s and the *G*s are equinumerous, i.e., there is a one-to-one correspondence between the *F*s and the *G*s:

$$\#F = \#G \leftrightarrow F \approx G$$

The analytic truth about directions and Hume's Principle are both examples of abstraction principles that give a criterion of identity for new concepts by 'carving up' previous propositional content. In Hume's Principle, content about equinumerosity is carved up to concern the concept of number. This is the principle behind Linnebo's notion of thin objects. Abstract objects like numbers and directions are referents of singular terms in the abstraction principles, and that is *all* they are required to be. Since they only feature in abstraction principles, their existence does not make any substantial demand of the world (Linnebo, 2018b; Pantsar, 2021c). The thin notion of mathematical objects requires no epistemological connection that is not present in the abstraction principles, which avoids Benacerraf's problem of establishing a connection between physical subjects with non-physical objects.

One may question, however, the sense in which thin objects are supposed to exist in a mind-independent manner, that is, just how *platonist* the thin objects account is. This is what Rayo's account of subtle platonism, which is closely related to Linnebo's idea of thin objects, aims to explain. Rayo's (2015) fundamental idea is that mathematical objects exist, but only in a *trivial* sense. This is spelled out in terms of compositional semantics which, Rayo argues, are trivial for mathematical statements. The problem is certainly important, as we have seen. According to standard semantics, existential mathematical statements can only be true if abstract mathematical objects exist. This has led many philosophers to paraphrase mathematical statements in nominalist or modal terms (Chihara, 1973, 1990; Hellman, 1989; Putnam, 1967). Instead of statements like 'there exists *x* such that . . .', the paraphrase can be, for example, 'it is possible to construct *x*' (Chihara, 2005). But, as Shapiro (1997, p. 228) has noted, there is something troubling about the idea of solving ontological problems by a change of vocabulary. From a similar background, Rayo (2015) argues against the paraphrasing approach to avoid existential quantification. Instead, he wants to show that we can stick to standard semantics because the truth-conditions for mathematical statements do not require substantial existence claims, that is, they are trivial.

Thus, Rayo (2013, 2015) accepts that arithmetical statements commit to numbers, but this commitment is to nothing non-trivial. In his example sentence 'the number of dinosaurs is zero', the truth conditions only require that there are no dinosaurs, that is, 'For the number of the dinosaurs to be zero *just is* for there to be no dinosaurs' (Rayo, 2015, p. 81, original emphasis). More generally, Rayo contends that there is no difference between there being n things that have the property F and the number of Fs being n: 'For the number of the Fs to be n just is for it to be the case that $\exists!_n x(Fx)$' (Rayo, 2015, p. 81, original emphasis).[4] Moving to purely arithmetical sentences, Rayo argues that the truth conditions are satisfied trivially regardless of what the world is like. For example, in the biconditional

$$\text{'}2 + 2 = 4\text{' is true if and only if } 2 + 2 = 4$$

the truth condition (the right-hand side) is fulfilled in all possible worlds. This way, Rayo's account aims to provide an alternative to paraphrasing or revisionist alterations of mathematics, without causing Benacerraf's epistemological problem about physical subjects having access to abstract objects. Subtle platonism takes mathematical practice as it is and maintains that a commitment to numbers need not carry any epistemological problems.

[4] The expression $\exists!_n x(Fx)$ means there are exactly n values of x for which it holds that Fx.

One potential problem in Rayo's account, like in that of Linnebo's thin objects, is that it is not immediately clear why it should count as platonist at all. In particular, what is there to prevent an alternative interpretation of it, according to which we are only committing to a particular linguistic practice, and not to the existence of numbers at all? And, if indeed all we have are linguistic practices, then what is there to distinguish subtle platonism or thin objects from conventionalism? Clearly it is a non-conventional fact that the number of dinosaurs is zero, but what prevents pure arithmetical statements like $2 + 2 = 4$ from being arbitrary conventions? Rayo's (2015) answer is that the truth conditions of $2 + 2 = 4$ are trivial because they are satisfied regardless of what the world is like. Hence, the existence of numbers is trivial, because there is no space for non-trivial truth-makers for arithmetical statements.

However, following my argument in Section 8.3, in the present approach it is hardly clear that the truth conditions of arithmetical statements are trivial. In possible worlds where there are no subjects that process observations in terms of numerosities, even a basic arithmetical statement like $2 + 2 = 4$ would not be trivial. After all, in such a possible world any statement involving natural numbers would require a new type of conceptual thinking. The proposition $2 + 2 = 4$ would not be *false* in such possible worlds, but there would be no cognitive agents who possess number concepts. For the truth conditions of $2 + 2 = 4$ to be satisfied, there have to be cognitive agents with number concepts, or some equivalent numerical ability. The existence of such cognitive agents, however, is not trivial.[5]

In a similar manner, I see the main problem in Linnebo's thin object account in the order of Hume's Principle and number concepts. In that account, content about equinumerosity is reconceptualised to concern numbers. As we have seen in Part I of this book, however, empirical data show that children acquire their first number concepts before they understand the principle of equinumerosity. Thus, Hume's Principle, in cognitive terms, is only possible to grasp once the subject already possesses some number concepts. Number concepts do not enter our cognitive abilities due to Hume's Principle and, consequently, I do not see any reason to

[5] One might still insist that the truth conditions of $2 + 2 = 4$ are trivial regardless of the existence of cognitive agents, it is only our *appreciation* of the truth conditions that is non-trivial. However, it is difficult to see what the agent-independent truth conditions could be, unless we are already subscribing to a platonist view. I thank Bahram Assadian for pointing out this potential problem.

believe that numbers exist primarily due to playing a referential role in Hume's Principle.

The upshot of this is that we do not need to think of thin objects as existing in any platonist sense – subtle or not. In this book, I have argued that natural numbers exist as the referents of culturally shared number concepts. Now we can see that this account is actually compatible with Linnebo's notion of thin objects, as long as we are not committed to the mind-independent existence of numbers. Like Linnebo, I am convinced that the existence of numbers is 'thin' in the sense that it makes no demands of the world. However, unlike Linnebo, I claim that we can only be sure of this thin existence occurring when there are cognitive agents who possess number concepts.[6]

It could be argued that this line of thinking does make one demand of the world, namely that there are sufficiently developed cognitive agents that possess number concepts (Pantsar, 2023d). However, this is hardly the kind of ontological commitment that Linnebo wants to avoid. Therefore, rather than seeing my account as being in conflict with Linnebo's, I propose the present epistemological theory as an alternative, non-platonist, interpretation of Linnebo's theory. Numbers as abstract objects do have a thin existence, but this refers to them existing only through culturally shared concepts, which makes abstraction principles like Hume's Principle possible. In Section 9.3, I will explain in detail what it means for numbers as objects to exist through culturally shared concepts.

9.3 Constructivism

According to the present account, neither is there any reason nor does it make sense to postulate that numbers exist as mind-independent entities. Obviously, the existence of numbers is not tied to any particular minds but, given that numbers exist as the referents of culturally shared concepts, the present account is committed to human minds playing a crucial role for the ontology of numbers. It is important to distinguish this position from *nominalist* views of mathematics. I am not claiming that numbers as abstract objects do not exist. Instead, I claim that their existence is the result of certain human cognitive practices. Therefore, the present position

[6] This does not imply that, say, $2 + 2 = 4$ was *false* before there were human beings or that it becomes false if humans (and other arithmetical agents) disappear. Rather, in the present account questions concerning such scenarios are not well placed.

is clearly *constructivist*: numbers as abstract objects are human constructions based on culturally shared cognitive practices.

The ontology of such abstract objects has been an important topic in philosophy. Popper (1972, chapter 3), for example, famously postulated a 'World 3' for the products of thought, to separate them from material and mental realms (Worlds 1 and 2, respectively). This World 3, the 'cultural realm', is clearly where numbers in the present account would be located in Popperian ontology. However, here I do not want to commit to Popper's ontological theory. In particular, I am reluctant to follow the general distinction of entities into different worlds in the manner of Popper. A full treatment of this topic is not possible here, but it is important to note that the account of numbers developed in this book is not limited to any particular theory concerning the existence of culturally developed entities. Indeed, I believe that it is compatible with most general forms of social constructivism.

Social constructivist accounts of mathematical objects are not new in the philosophy of mathematics. The most famous such accounts have been presented by Julian Cole (2013, 2015) and Solomon Feferman (2009). Cole has argued that 'mathematical entities are pure constitutive social constructs constituted by mathematical activities' (Cole, 2009, p. 599, emphases removed). More specifically, based on Searle's (1997, 2010) theory of the construction of social reality, Cole argues that the existence of mathematical objects is the result of imposing *institutional* functions onto reality (Cole, 2013, 2015). That is how, for example, a piece of paper becomes currency. Institutional impositions depend on there being constitutive rules for their existence, which is how laws, rules of games and so on come to existence. One important class of institutional functions are *representational* functions, which 'allow us to represent the world using intentional states that structure it into entities with features' (Cole, 2013, p. 14). Mathematical objects, then, are institutional entities that play a representational function. Their 'function is to facilitate our abilities to represent, analyse, reason about, discover truths concerning, etc. facets of reality that are not the entities in question' (Cole, 2013, pp. 13–14).

In short, Cole's theory is that we introduce entities with representational functions, such as natural numbers, because it is in some way cognitively advantageous to handle a representation of reality in terms of numbers. Numbers themselves are social constructions, institutionally imposed, but through their representational functions they are connected to the world around us. Feferman's (2009) social constructivist account is fundamentally similar to Cole's. According to Feferman, 'mathematics emerges from humanly constructed, inter-subjectively established, basic structural conceptions' (Feferman, 2009, p. 1). Feferman presents a list of ten theses concerning how this emergence happens, of which two are important to mention here:

> The basic objects of mathematical thought exist only as mental conceptions, though the source of these conceptions lies in everyday experience in manifold ways, in the processes of counting, ordering, matching, combining, separating, and locating in space and time.
>
> . . .
>
> The basic conceptions of mathematics are of certain kinds of relatively simple ideal-world pictures which are not of objects in isolation but of structures, i.e. coherently conceived groups of objects interconnected by a few simple relations and operations. They are communicated and understood prior to any axiomatics, indeed prior to any systematic logical development. (Feferman, 2009, p. 3)

I believe that Feferman's account is highly compatible with the epistemological account I have developed in this book. I have also supported the view that the basic objects of arithmetic (i.e., natural numbers) exist as mental concepts, although I have argued for the weaker position that they exist *primarily*, not necessarily *only* as such.[7] I have supported the view that the basic concepts of arithmetic are relatively simple ideal-world pictures, communicated and understood prior to any axiomatics or systematic logic. Indeed, I see my position as fundamentally in line with the social constructivism of Cole and Feferman. However, I believe that the present account is considerably stronger than either of those accounts. That is because, unlike them, I have provided reasons *why* mathematical concepts like numbers play successful representational functions (unlike Cole) and why their source lies in everyday experience (unlike Feferman). The reason for this is, as has become clear by now, that numbers are social constructs based on our *proto-arithmetical* abilities. I will develop this point further next.

One great strength of the present account over those of Cole and Feferman is how it can explain mathematical objectivity. For Cole, the success of numbers as social constructs is explained through their representational function. In particular, he argues that the cognitive constitution of human beings makes it easier for us to represent the world in terms of numbers (Cole, 2013, p. 14). Because of this success of numbers,

[7] Feferman writes about 'conceptions' but I believe this to be synonymous to my use of 'concepts'.

they are institutionalised as social constructs. I find this generally a very plausible position. Numbers become part of institutions in science, education, commerce and other fields of societal activity because their representational function is closely connected to the human cognitive constitution. However, unlike Cole, I have provided an explanation why this is the case. It is because an important part of our cognitive constitution is observing our surroundings in terms of discrete macro-level objects through our proto-arithmetical abilities. Since these abilities are universally shared, it gives rise to a strong level of objectivity.[8]

While I believe I have explained the objectivity of arithmetic in a satisfactory manner, one last question remains: how exactly do social constructs like numbers exist as objects? I have argued that numbers as abstract objects exist as the referents of culturally shared concepts, but this requires further analysis. The question is closely related to a central ontological question in the social constructivist accounts, namely, how social constructs generally exist. The short answer is that this happens through human interactions, but this does not help much ontologically. In the specific case of natural numbers, human interactions cause us to possess shared number concepts. But how do we come to *objects* from those concepts?

This might seem like a difficult philosophical question, and perhaps from the perspective of general ontology it is. However, from the perspective of this book, the question is not particularly problematic. I have argued for a particular account of epistemology of arithmetic, in which arithmetic develops based on proto-arithmetical abilities through processes of enculturation (in the individual ontogeny) and cumulative cultural evolution (in phylogeny and cultural history). In this development, members of cultures acquire natural number concepts. On the personal level, arithmetical cognition is about those number concepts and their physical and verbal representations (i.e., numeral words and symbols). Described in this way, numbers as objects are not needed to make sense of this account. But, in the account that I have presented, it is integral that the number concepts are *shared* within (but also across) cultures. For this purpose, it is important that we can speak of arithmetic without depending on particular instantiations of numeral words, numeral symbols and number concepts. In short, for successful arithmetical practices, we need to make sure that my number concepts and arithmetical beliefs mean the

[8] For a more detailed discussion of social constructivism related to the present epistemological theory, see Pantsar (2023b).

same as yours. To make this possible, we introduce numbers as social constructs. We start treating numbers as objects, which allows us to determine that we indeed possess similar number concepts.

All this is likely to have happened implicitly, without there ever being any declaration of numbers as objects. This most probably simply took place as a by-product when symbolic and verbal representations developed concurrently with number concepts, as described in Part II of this book. Perhaps this happened simply to facilitate communication: instead of always talking about the process of counting, it was easier to enter discourse on numbers as objects. This move can be explained by the Process → Object Metaphor, as done in Section 5.1 (Pantsar, 2015b). We cannot trace how this development happened in the Western cultural trajectory, but we do know that it ended up in discourse that treats natural numbers as abstract objects. We could take the nominalist approach and deny that they are such. However, given all that we have seen in this book, this seems misguided. Why not rather try to explain what *kind* of objects natural numbers are? Here I have proposed an answer to that: natural numbers as abstract objects are social constructs in the particular sense of referents of culturally shared number concepts. The only ontological accounts that my answer here contradicts are ones that deny the existence of referents of shared concepts. While this kind of view is impossible to refute, such a solipsistic notion of concepts is not particularly interesting. If we accept the existence of shared concepts, I believe we can develop an account of abstract objects based on them, as I have done in this book.

In developing this account, I may be accused of disregarding Frege's (1892) famous distinction between object and concept, but it has to be remembered that my approach is entirely different from Frege's. In the present approach, concepts are primarily mental entities and it is not a trivial matter that concepts are shared across individuals. This is definitely the case when those concepts do not refer to mind-independent entities. Perhaps a different word could be used for the referents of such shared, mind-dependent concepts related to social constructs, but I believe that 'abstract object' is as good as any.

One final question remains: what does it mean for objects to be the referents of culturally shared number concepts? Given that numbers are not thought to have any ontological status beyond being the referents of shared number concepts, this question gets a straightforward answer. The referents are the social constructs that the number concepts are associated with. These referents are abstract because they are not located in time or space. However, that does not imply that they have always existed.

Numbers, according to the present account, are human constructions and as such they did not exist before there were human arithmetical cultures. In this sense, they are like any other abstract social constructs.

9.4 Ordinal or Cardinal

To end this chapter, I will tackle a topic that has traditionally been seen as important in the philosophy of arithmetic. In the introduction, it was seen that natural numbers can be thought of both as cardinals and ordinals. In the philosophy of mathematics, this gives rise to an old and venerable ontological problem, namely, whether natural numbers exist *primarily* as ordinals or cardinals. For finite non-zero natural numbers, this does not make a difference arithmetically, since there is a simple way to connect the ordinal and cardinal interpretation. In the standard counting list, 'one' is the first numeral word, 'two' is the second numeral word and so on. For infinite sets of natural numbers, the matter is quite different, but here it is not possible to go into that topic. However, the double role of natural numbers is also relevant when considering numerical cognition and number concept acquisition. The obvious question to ask is whether grasping natural numbers as cardinals is primary to grasping them as ordinals. This question is closely related to an old discussion in the foundations of mathematics, that is, whether natural numbers are *fundamentally* cardinals or ordinals.

The origin of this topic traces back to Cantor (1883), who was the first to define cardinal numbers based on ordinal numbers in the well-ordering of natural numbers. Cantor's position has generally been interpreted as implying the primariness of ordinal numbers over cardinal numbers (Assadian & Buijsman, 2019; Hallett, 1988). Similarly, in the set theoretic definition of natural numbers, ordinals are often seen as more fundamental than cardinals. In the Von Neumann construction, natural numbers are defined by the power set operation so that 0 is the empty set $\emptyset$, 1 is the set formed by the empty set $\{\emptyset\}$, 2 is the set formed by 0 and 1, that is, $\{\emptyset, \{\emptyset\}\}$ and so on (Enderton, 1977). The cardinal number of a set can then be defined as the (least) ordinal number whose members can be put in one-to-one correspondence with the members of the set (Assadian & Buijsman, 2019, p. 565). For finite numbers, this number is of course unique: a set with, say, five members can be put in one-to-one correspondence with the ordinal number five.

In mathematical structuralism, as explained in Section 9.1, mathematical objects exist as places in mathematical structures. Thus, the number five, for example, exists fundamentally as its place in the natural number structure. This implies that the fundamental existence of numbers, whatever it may be in terms of ontology, is as ordinal numbers. In the *ante rem* platonist account of structuralism of Shapiro, for example, natural numbers are explicitly defined in terms of their (ordinal) position in the natural number structure (Shapiro, 1997, p. 72).

However, Shapiro's structuralism and the resulting primariness of ordinal numbers is not characterised in terms of either the ontogenetic cognitive development of number concept acquisition or the phylogenetic and historical development of number concepts within cultures. In the present account, we have made a commitment to look for cognitively fundamental notions. There is little hope of determining whether ordinals or cardinals came first in the historic development, but the ontogenetic question can be tackled. Thus, the important question here is whether children's first access to natural number concepts is through their cardinal or ordinal character. Given the importance of counting lists for the bootstrapping account, it might seem clear that the present approach is committed to the primariness of cardinality of natural numbers. However, transitive counting by itself could also be an ordinal process. Just like we can count 'one, two, three, four bottles', we can count 'first, second, third, fourth bottle'. Therefore, the role of counting lists alone does not imply the primariness of either the ordinal or the cardinal conception.

However, the role of the OTS and the ANS in the present account definitely suggests that cardinality is the primary access to natural numbers. Both proto-arithmetical abilities, subitising and estimating, detect the cardinality of the observed collection. There is nothing to suggest, for example, that the order of the occupied object files in the OTS has significance for the subitising ability. Therefore, for the acquisition of the first number concepts at least, the present account implies that cardinality is primary. This is in line with empirical studies, which show that children who lack numeral words cannot make ordinal judgments (see, e.g., Brannon & Van de Walle, 2001). Given that non-verbal children can make cardinality judgments on numerosity, it seems that the ordinal understanding is something that only emerges with numeral words.

Indeed, it seems plausible that the cardinal understanding is at the root of acquiring the very first number concept. Buijsman (2019) has suggested that the acquisition of the number concept ONE depends on understanding claims of the logical form $\exists! xFx$ (i.e., there exists exactly one *F*). To understand such claims, one needs to use syntactic clues to grasp the singular/plural distinction, and thus grasp the correct usage of the word 'one' (Buijsman, 2019; Syrett et al., 2012). This is coherent with the present account and could give us a better understanding of early number concept acquisition. Just like proto-arithmetical abilities, the syntactical singular/plural distinction does not concern ordinality.

More data is definitely needed before any firm claims about the priority of cardinality or ordinality can be made. But at the very least the above considerations suggest that the acquisition of a cardinal understanding of natural number concepts can be independent of their ordinal understanding.[9] This would suggest that the primariness of ordinality in the structuralist account, for example, does not capture a necessary order of cognitive development in number concept acquisition. Indeed, the current understanding of the topic points to the opposite view: that children's first grasp of number concepts is cardinal. This does not mean that the ordinal understanding could not also be important in number concept acquisition. The principle of well-ordering, for example, is implicitly present in the counting games that are thought to play an important role in the bootstrapping process (Beck, 2017). Thus, distinguishing between the ordinal and cardinal elements may not ultimately be a simple task.

What does all this mean for the ontology of natural numbers, that is, do natural numbers exist fundamentally as cardinal or ordinal numbers? Given that, in the present account, natural numbers are seen to exist as referents of culturally shared concepts, the epistemology of ordinality and cardinality is clearly relevant also for the ontological question. If epistemological considerations point to the primariness of cardinality, this means that we should consider natural numbers to be fundamentally cardinal. As of now, however, there is not enough knowledge about number concept acquisition to make any conclusive statements. Nevertheless, I hope that the present account at the very least provides a strong argument against the notion held by Cantor and Shapiro, among others, that natural numbers must be fundamentally ordinal.

9.5 Summary

In this chapter, I moved the focus from epistemology to ontology. I asked what kind of objects natural numbers are and showed why different types of realist answers are problematic. I then endorsed a constructivist account,

[9] Although Buijsman (2021) has argued that, when one gets to larger natural numbers, the ordinal conception becomes more important.

according to which natural numbers exist as social constructs through our shared number concepts. This account is compatible with the social constructivist views of Cole and Feferman. However, I see my account as an improvement on their work because it provides a strong explanation of *why* numbers are widely applicable social constructs. This explanation is based on the important role that proto-arithmetical abilities have both for the way we experience our environment and the development of arithmetic. Finally, I discussed the question of whether natural numbers as social constructs exist primarily as ordinals or cardinals, showing that, although more empirical research is needed, there is no reason to currently believe that natural numbers must be fundamentally ordinal.

Conclusion

In the Introduction, I presented the question of how arithmetical knowledge can be acquired as the most fundamental question in the epistemology arithmetic. In Part I, I have presented an answer to this question in terms of individual ontogeny. Arithmetical knowledge can be acquired by individuals through a process of enculturation in which they employ their proto-arithmetical abilities. However, this answer is incomplete. I noted that, in addition to ontogeny, we need an account of the development of arithmetical knowledge in terms of phylogeny and cultural history. In Part II, I provided an outline of such an account. Arithmetical knowledge develops in cultures through a process of cumulative cultural evolution, in which material engagement with the environment plays a key role. Also, in this process, the shared proto-arithmetical abilities are crucial. Then, in Part III, I asked some fundamental philosophical questions concerning the account I have proposed. Perhaps most importantly, I wanted to distinguish my account from conventionalism. I argued that, due to the proto-arithmetical origins of arithmetic, we can save objectivity of arithmetic in a strong enough sense to avoid the threat of conventionalism. In addition, I argued that we can save the apriority, necessity and universality of arithmetic, although all of them in a different sense than in traditional platonist philosophy.

As mentioned in the Preface, this book can be seen as the culmination of a long project. I started with this research topic in 2010 and, at that point, I had only a very vague understanding of the vastness of the project. However, in the beginning I already had a clear idea of what criteria the emerging epistemological theory should fulfil. I was dissatisfied with the epistemological accounts of arithmetic presented in the literature and I wanted to make sure that, whatever the account I end up endorsing may be like, it would not fall into the same pitfalls. For this reason, in the first major published work of the project (Pantsar, 2014), I proposed a wish-list for an epistemological account of arithmetic. During the long process that has followed, what started as early ideas have become clearer and more definite. Meanwhile, empirical research on numerical cognition has developed a great deal since the spring of 2012, when the first paper was originally submitted. However, nothing has changed in my attitude toward the criteria I presented. Therefore, it is good to return to the wish-list in this concluding section.

The five wishes I identified for an epistemological account of arithmetic were the following:

> 1. It should not require any unreasonable ontological assumptions.
> 2. It should be epistemologically feasible as part of a generally empiricist philosophy.
> 3. It should be able to explain the apparent objectivity of at least some mathematical truths.
> 4. It should not make the applications of mathematical theories in empirical sciences a miracle.
> 5. It should not rid mathematics of its special character
>
> (Pantsar, 2014, p. 4202).

I believe that the present account fulfils all five wishes. First, no unreasonable ontological assumptions are made. Numbers are understood to exist as social constructs, referents of culturally shared concepts, and no other mode of existence for them is assumed. Thus, the present account only requires that there exist cultures and concepts can be shared within them. While there may be disagreement over the exact definitions and scopes of the notion's 'culture' and 'concept', ultimately the present account is flexible in that regard.

The second wish is also clearly fulfilled. In building the epistemological account, no special epistemological faculty related to arithmetic has been suggested. The only kind of empiricist epistemology that my account conflicts with is a 'blank slate' understanding of the mind. However, such theories have been widely discredited in the modern study of cognition (see, e.g., Dehaene, 2020). As for the third wish, the argumentation in Section 8.2 shows how arithmetical knowledge based on proto-arithmetical abilities is likely to appear as objective to us. We cannot change our basic cognitive architecture, so we cannot have arithmetic that conflicts with proto-arithmetical abilities. While this does not mean that arithmetic *is* strictly objective, at least in its entirety, it explains why arithmetic *appears* to be objective.

The fourth wish is a topic that I have not discussed extensively in this book so far. However, it is a particularly important topic because I have argued that, in addition to the apparent objectivity, scientific applications are the main reason for believing in the objectivity of mathematics (Pantsar, 2021c). Indeed, it has been used as the main argument for objectivity in the recent philosophical literature (e.g., Brown, 2008; Colyvan, 2001; Lange, 2017). The existence of mathematical applications outside mathematics is problematic both for strict conventionalist accounts and platonist accounts. For the former, the problem is how ultimately (possibly) arbitrary human conventions can help explain the world. For the latter, the problem is how the platonist world of abstract ideas is connected to the physical world. The present account, however, carries no such problems. Contra conventionalism, arithmetic cannot be understood as arbitrary; rather, its content is constrained by our basic cognitive architecture. Contra platonism, arithmetic is not understood to concern a platonist world of ideas; it ultimately concerns our basic ways of observing the world. Thus, arithmetical knowledge is tightly connected to our basic cognitive capacities. It could be that our cognitive capacities have evolved to detect the structure of the world, thus connecting mathematics directly to the physical sciences whose aim is to explain that structure, as argued by Maddy (2014) and De Cruz (2016). But we do not need to make this assumption. It could also be the case, as I suggested in Section 8.2, that our cognitive capacities evolved simply to enhance survival and reproduction. Also, in this case, however, arithmetic is tightly connected to the physical sciences. In the sciences, we are ultimately explaining the world through our observations. If our observations are shaped by proto-arithmetical abilities, it is no miracle that arithmetic has applications in the sciences. Arithmetic is a systematic development based on proto-arithmetical abilities, which shape our observations in important ways. It is thus to be expected that arithmetic (and by extension other mathematical theories) have applications in scientific theories.

Finally, the fifth wish is fulfilled in the present account due to the contextual a priori character of arithmetical knowledge. Within general empiricist epistemology, mathematical knowledge clearly has a special role, and the account developed in this book manages to explain that. Arithmetical knowledge is neither refuted nor corroborated by observations, because it is a priori in the context set by our proto-arithmetical abilities. This is an important advancement over empiricist theories like those of Mill (1843) and Kitcher (1983). Those theories ran into problems because they did not include a special character for mathematical knowledge. But I do believe that mathematical knowledge is special. It is not simply empirical knowledge based on extremely wide generalisations.

Consequently, I maintain that an epistemological account of mathematics must be able to explain this special character. In this book, I focused on three such special characteristics: apriority, universality and necessity. In Chapter 8, I argued that all characteristics can be explained by the present account. Granted, we needed to tweak our understanding of the relevant notions. But, as I have shown, the account of arithmetical knowledge as contextually a priori, based on proto-arithmetical abilities, can explain each characteristic in a sufficiently strong form.

One question that the reader may have wondered about is whether arithmetic is different from other fields of mathematics in its cognitive origins. In the project of Lakoff and Núñez (2000), for example, the scope was not limited to arithmetic. Is my project similarly extendable to other areas of mathematics? My hypothesis is that it is, but this should not be taken as a foregone conclusion. I cannot imagine how other fields of mathematics could have developed essentially differently, that is, not through enculturation and cumulative cultural evolution. Neither can I imagine how other mathematical objects could exist in a different manner from natural numbers. But these are simply impressions that I have. I believe in the value of the kind of work that I have reviewed and contributed to in this book. The cognitive and cultural details matter, and we cannot simply assume that the details can be filled out. So far the research focus has been mainly on arithmetic, but there have also been similar developments to explain the development of Euclidean geometry based on proto-geometrical abilities (Hohol, 2019; Pantsar, 2021d). In the future, I hope that other fields of mathematics will get more attention and that the work done on arithmetical cognition will prove to be helpful in that.

Indeed, aside from the main purpose of presenting my epistemological account of arithmetic, one aim of this book has been to introduce conceptual tools for critically assessing empirical research on numerical cognition. In this, I hope that there are wider lessons to be learned for the study of mathematical and proto-mathematical cognition. One important lesson concerns the danger of confusing the two types of cognition. While I believe that there is plenty of evidence to suggest that arithmetic is based on proto-arithmetical abilities, I also want to stress that many empirical results are reported in highly misleading ways. Numerical cognition is an active research topic and most likely we will hear much exciting news in the field in the coming years. Probably we will see many headlines like 'Math Bee: Honeybees Seem to Understand the Notion of Zero' (Greenfieldboyce, 2018). One thing I hope that this book has managed to do is to make the reader sceptical of such claims. Simply put, there is no evidence whatsoever that bees understand the notion of zero. What they could do in the reported experiment is determine (not very accurately but above chance) that a card with no symbols represents 'less than' a card with symbols (Howard et al., 2018). This is of course remarkable on its own, and an exciting example of how insect intelligence may exceed the levels we have traditionally granted it. Bees, it seems, have an ability to grasp the *absence* of something as significant. That is no small cognitive feat. Similar abilities have been detected also in other animals, including chimpanzees (Biro & Matsuzawa, 2001) and parrots (Pepperberg & Gordon, 2005), but it truly is a remarkable finding if also insects can in some way grasp nothingness.

Nevertheless, it is important to remember that zero is not the same as absence or nothingness. The way the bee experiment was reported in *Science* starts as follows: 'Humans' invention of zero was crucial for modern mathematics and science, but we're not the only species to consider "nothing" a number. Parrots and monkeys understand the concept of zero, and now bees have joined the club, too' (Warren, 2018). As the reader will immediately recognise, many things go wrong here. First of all, to the best of our knowledge, parrots and monkeys do not understand the concept of zero, simply because they do not possess number concepts.[1] We are the only species, as far as we know, to consider 'nothing' a number, simply because we are the only species that considers *anything* to be numbers. But, even allowing for these terminological problems, there is one particularly troubling aspect to the above quotation: it puts the human invention of zero parallel to the animal ability to grasp absence or nothingness. It is true that the invention of zero was very important for modern mathematics and science, and it is remarkable that the mathematical concept of zero only reached Western Europe in the twelfth century (Merzbach & Boyer, 2011). It is of course also true that the lack of zero did not prevent the Ancient Greek mathematicians from making great progress. But, generally, mathematics and science greatly benefitted from the introduction of the mathematical concept of zero.

Then, one must wonder, how is it possible that Europeans only acquired the concept of zero in the twelfth century when even parrots,

[1] This idea of animal possessing the concept of zero is quite common, but many researchers soften it by saying, for example, that animals have 'zero-like' concepts (see, e.g., Nieder, 2019, p. 295). I agree that animals can observe the absence of things, and perhaps some animals develop some kind of notion of 'nothing' based on that. It is controversial whether this can be considered 'zero-like' in that it is treated as a quantity.

monkeys and bees have it? The absurdity of this question should show clearly what the most important problem in the above quotation from *Science* is. The Ancient Greek people understood the notion of nothing, they just did not have a mathematical symbol for it. It was not the notion of nothing that made it possible to have new advances in mathematics and science; it was the innovation of being able to conceptualise nothing exactly and present it symbolically. But, of course, this latter is beyond the abilities of monkeys, parrots and bees. What they can have is a *proto-arithmetic* grasp of nothing that relates to numerosities or is perhaps even treated as a numerosity itself.[2]

On that note, I want to leave the reader with both an optimistic and a cautious mindset. What I have presented in this book is based on the present state of the art in the empirical study of numerical cognition. Given the fast development in the field, it is almost certain that a great amount of new relevant data will emerge in the next years. Thus, this book, at least in its reporting of the empirical studies, may be outdated faster than I would like. However, I am hopeful that future data can be assimilated in the present epistemological account. As our empirical understanding of the development of numerical cognition develops, the stronger this account will most likely also become. If conflicting data emerges, however, I would welcome that as well. For example, if it turned out that non-human animals possess number concepts, or that humans have innate number concepts, it would be a great step forward in explaining the development of arithmetical cognition. Both scenarios seem extremely unlikely to me, but we should definitely remain open to their possibility. But before we start changing our accounts of the development of numerical and arithmetical cognition based on new data, we need to be clear what the data are *about*. This is the cautious message I want to end with. Honeybees possessing number concepts may make for exciting headlines, but we should not be fooled by flashy news reports. In order to avoid confusions, we need to understand clearly what is being discussed, from the proto-arithmetical origins to formal theories of arithmetic. I hope this book will be able to make a positive contribution in that.

[2] In fact, based on the experiment reported in Howard et al. (2018), this remains undecided. While the bees seemed to recognise that no symbol on a card means less than one, two or three symbols, we cannot know that they grasped that no symbol on a card means nothing. It could also represent other notions of 'less than one', such as a fraction.

Bibliography

Adamson, L. B. & Bakeman, R. (1991). The development of shared attention during infancy. In R. Vasta (Ed.). *Annals of Child Development*, Vol. 8. (pp. 1–41). Jessica Kingsley.

Agrillo, C. (2015). Numerical and arithmetic abilities in non-primate species. In R. C. Cohen Kadosh & A. Dowker (Eds.), *The Oxford Handbook of Numerical Cognition* (pp. 214–236). Oxford University Press.

Amalric, M. & Dehaene, S. (2016). Origins of the brain networks for advanced mathematics in expert mathematicians. *Proceedings of the National Academy of Sciences, 113*(18), 4909–4917.

Amalric, M., Denghien, I. & Dehaene, S. (2018). On the role of visual experience in mathematical development: Evidence from blind mathematicians. *Developmental Cognitive Neuroscience, 30*, 314–323. https://doi.org/10.1016/j.dcn.2017.09.007.

Anderson, M. L. (2010). Neural reuse: A fundamental organizational principle of the brain. *Behavioral and Brain Sciences, 33*(4), 245–266.

(2015). *After Phrenology: Neural Reuse and the Interactive Brain*. MIT Press.

(2016). Précis of after phrenology: Neural reuse and the interactive brain. *Behavioral and Brain Sciences, 39*, e120. https://doi.org/10.1017/S0140525X15000631.

Ansari, D. (2008). Effects of development and enculturation on number representation in the brain. *Nature Reviews Neuroscience, 9*(4), 278–291.

(2012). Culture and education: New frontiers in brain plasticity. *Trends in Cognitive Sciences, 16*(2), 93–95.

Ashkenazi, S., Henik, A., Ifergane, G. & Shelef, I. (2008). Basic numerical processing in left intraparietal sulcus (IPS) acalculia. *Cortex; a Journal Devoted to the Study of the Nervous System and Behavior, 44*(4), 439–448. https://doi.org/10.1016/j.cortex.2007.08.008.

Assadian, B. & Buijsman, S. (2019). Are the natural numbers fundamentally ordinals? *Philosophy and Phenomenological Research, 99*(3), 564–580. https://doi.org/10.1111/phpr.12499.

Ayer, A. J. (1970). *Language, Truth and Logic*, 2nd ed. Dover.

Aziz, T. A., Pramudiani, P. & Purnomo, Y. W. (2017). How do college students solve logarithm questions? *International Journal on Emerging Mathematics Education, 1*(1), 25–40.

Balaguer, M. (2016). Platonism in metaphysics. In E. N. Zalta (Ed.), *The Stanford Encyclopedia of Philosophy*. Metaphysics Research Lab, Stanford University. https://plato.stanford.edu/archives/spr2016/entries/platonism/.

Barrocas, R., Roesch, S., Gawrilow, C. & Moeller, K. (2020). Putting a finger on numerical development: Reviewing the contributions of kindergarten finger gnosis and fine motor skills to numerical abilities. *Frontiers in Psychology*, *11*, 1012. https://doi.org/10.3389/fpsyg.2020.01012.

Beck, J. (2017). Can bootstrapping explain concept learning? *Cognition*, *158*, 110–121.

Benacerraf, P. (1965). What numbers could not be. *The Philosophical Review*, *74*(1), 47–73. https://doi.org/10.2307/2183530.

(1973). Mathematical truth. *Journal of Philosophy*, *70*, 661–679.

Bender, A. & Beller, S. (2012). Nature and culture of finger counting: Diversity and representational effects of an embodied cognitive tool. *Cognition*, *124*(2), 156–182. https://doi.org/10.1016/j.cognition.2012.05.005.

Ben-Menahem, Y. (1998). Explanation and description: Wittgenstein on convention. *Synthese*, *115*(1), 99–130.

Bennett, M. R. & Hacker, P. M. S. (2003). *Philosophical Foundations of Neuroscience*. Blackwell.

Beran, M. J., Evans, T. A., Leighty, K. A., Harris, E. H. & Rice, D. (2008). Summation and quantity judgments of sequentially presented sets by capuchin monkeys (*Cebus apella*). *American Journal of Primatology*, *70*(2), 191–194. https://doi.org/10.1002/ajp.20474.

Berlyne, D. E. (1966). Curiosity and exploration. *Science*, *153*(3731), 25–33. https://doi.org/10.1126/science.153.3731.25.

Bigelow, J. (1988). *The Reality of Numbers: A Physicalist's Philosophy of Mathematics*. Oxford University Press.

Biro, D. & Matsuzawa, T. (2001). Use of numerical symbols by the chimpanzee (*Pan troglodytes*): Cardinals, ordinals, and the introduction of zero. *Animal Cognition*, *4*(3–4), 193–199. https://doi.org/10.1007/s100710100086.

Bock, A. S., Binda, P., Benson, N. C., Bridge, H., Watkins, K. E. & Fine, I. (2015). Resting-state retinotopic organization in the absence of retinal input and visual experience. *Journal of Neuroscience: The Official Journal of the Society for Neuroscience*, *35*(36), 12366–12382. https://doi.org/10.1523/JNEUROSCI.4715-14.2015.

Boghossian, P. A. (1997). Analyticity. In B. Hale & C. Wright (Eds.), *A Companion to the Philosophy of Language* (pp. 331–368). Blackwell.

Bogoshi, J., Naidoo, K. & Webb, J. (1987). The oldest mathematical artefact. *The Mathematical Gazette*, *71*(458), 294. https://doi.org/10.2307/3617049.

Boolos, G. (1998). *Logic, Logic and Logic* (R. Jeffrey & J. P. Burgess, Eds.). Harvard University Press.

Bouhali, F., Thiebaut de Schotten, M., Pinel, P., Poupon, C., Mangin, J.-F., Dehaene, S. & Cohen, L. (2014). Anatomical connections of the visual word form area. *Journal of Neuroscience: The Official Journal of the Society for Neuroscience*, *34*(46), 15402–15414. https://doi.org/10.1523/JNEUROSCI.4918-13.2014.

Boyd, R. & Richerson, P. J. (1985). *Culture and the Evolutionary Process.* University of Chicago Press.

(1996). Why culture is common, but cultural evolution is rare. In W. G. Runciman, J. M. Smith & R. I. M. Dunbar (Eds.). *Evolution of Social Behaviour Patterns in Primates and Man* (pp. 77–93). Oxford University Press.

(2005). *Not by Genes Alone.* University of Chicago Press.

Boyer, C. (1991). *A History of Mathematics*, 2nd ed. John Wiley & Sons.

Boysen, S. T. & Berntson, G. G. (1989). Numerical competence in a chimpanzee (*Pan troglodytes*). *Journal of Comparative Psychology (Washington, D.C.: 1983), 103*(1), 23–31. https://doi.org/10.1037/0735-7036.103.1.23.

Brannon, E. M. & Van de Walle, G. A. (2001). The development of ordinal numerical competence in young children. *Cognitive Psychology, 43*(1), 53–81. https://doi.org/10.1006/cogp.2001.0756.

Bremner, J. G., Slater, A. M., Hayes, R. A., Mason, U. C., Murphy, C., Spring, J., Draper, L., Gaskell, D. & Johnson, S. P. (2017). Young infants' visual fixation patterns in addition and subtraction tasks support an object tracking account. *Journal of Experimental Child Psychology, 162*, 199–208. https://doi.org/10.1016/j.jecp.2017.05.007.

Brown, J. R. (2008). *Philosophy of Mathematics: A Contemporary Introduction to the World of Proofs and Pictures*, 2nd ed. Routledge.

Buijsman, S. (2019). Learning the natural numbers as a child. *Noûs, 53*(1), 3–22.

(2021). How do we semantically individuate natural numbers? *Philosophia Mathematica, 29*(2), 214–233. https://doi.org/10.1093/philmat/nkab001.

Buijsman, S. & Pantsar, M. (2020). Complexity of mental integer addition. *Journal of Numerical Cognition, 6*(1), 148–163.

Butterworth, B. (1999). *What Counts: How Every Brain Is Hardwired for Math.* Free Press.

(2005). The development of arithmetical abilities. *Journal of Child Psychology and Psychiatry, and Allied Disciplines, 46*(1), 3–18. https://doi.org/10.1111/j.1469-7610.2004.00374.x.

(2010). Foundational numerical capacities and the origins of dyscalculia. *Trends in Cognitive Sciences, 14*(12), 534–541. https://doi.org/10.1016/j.tics.2010.09.007.

Campbell, J. I. D. (1994). Architectures for numerical cognition. *Cognition, 53*(1), 1–44. https://doi.org/10.1016/0010-0277(94)90075-2.

Campbell, J. I. D. & Epp, L. J. (2004). An encoding-complex approach to numerical cognition in Chinese–English bilinguals. *Canadian Journal of Experimental Psychology = Revue Canadienne De Psychologie Experimentale, 58*(4), 229–244. https://doi.org/10.1037/h0087447.

Cantlon, J. F. & Brannon, E. M. (2007). Basic math in monkeys and college students. *PLoS Biology, 5*(12), e328. https://doi.org/10.1371/journal.pbio.0050328.

Cantlon, J. F., Merritt, D. J. & Brannon, E. M. (2016). Monkeys display classic signatures of human symbolic arithmetic. *Animal Cognition, 19*(2), 405–415. https://doi.org/10.1007/s10071–015-0942-5.

Cantor, G. (1883). Über unendliche, lineare Punktmannigfaltigkeiten, 5. *Mathematische Annalen*, *21*, 545–586.
Cantrell, L. & Smith, L. B. (2013). Open questions and a proposal: A critical review of the evidence on infant numerical abilities. *Cognition*, *128*(3), 331–352. https://doi.org/10.1016/j.cognition.2013.04.008.
Carey, S. (2009). *The Origin of Concepts*. Oxford University Press.
Carnap, R. (1937). *The Logical Syntax of Language*. Open Court.
Carruthers, P. (2006). *The Architecture of the Mind*. Clarendon Press.
Casasanto, D. & Boroditsky, L. (2008). Time in the mind: Using space to think about time. *Cognition*, *106*(2), 579–593. https://doi.org/10.1016/j.cognition.2007.03.004.
Castaldi, E., Pomè, A., Cicchini, G. M., Burr, D. & Binda, P. (2021). The pupil responds spontaneously to perceived numerosity. *Nature Communications*, *12*(1), 5944. https://doi.org/10.1038/s41467–021-26261-4.
Casullo, A. (2003). *A Priori Justification*. Oxford University Press.
Chalmers, D. J. (1997). *The Conscious Mind: In Search of a Fundamental Theory*, revised ed. Oxford University Press.
Chang, Y. (2014). Reorganization and plastic changes of the human brain associated with skill learning and expertise. *Frontiers in Human Neuroscience*, *8*, 35. https://doi.org/10.3389/fnhum.2014.00035.
Charette, F. (2012). The logical Greek versus the imaginative Oriental: On the historiography of 'non-Western' mathematics during the period 1820–1920. In K. Chemla (Ed.), *The History of Mathematical Proof in Ancient Traditions* (pp. 274–293). Cambridge University Press.
Chemla, K. (Ed.). (2015). *The History of Mathematical Proof in Ancient Traditions*, reprint ed. Cambridge University Press.
Cheung, P. & Le Corre, M. (2018). Parallel individuation supports numerical comparisons in preschoolers. *Journal of Numerical Cognition*, *4*(2), 380–409.
Cheyette, S. J. & Piantadosi, S. T. (2020). A unified account of numerosity perception. *Nature Human Behaviour*, *4*(12), 1265–1272. https://doi.org/10.1038/s41562–020-00946-0.
Chihara, C. (1973). *Ontology and the Vicious Circle Principle*. Cornell University Press.
(1990). *Constructibility and Mathematical Existence*. Oxford University Press.
(2005). Nominalism. In S. Shapiro (Ed.), *The Oxford Handbook of Philosophy of Mathematics and Logic* (pp. 483–514). Oxford University Press.
Chomsky, N. (2006). *Language and Mind*, 3rd ed. Cambridge University Press.
Chrisomalis, S. (2010). *Numerical Notation: A Comparative History*. Cambridge University Press.
Christodoulou, J., Lac, A. & Moore, D. S. (2017). Babies and math: A meta-analysis of infants' simple arithmetic competence. *Developmental Psychology*, *53*(8), 1405–1417. https://doi.org/10.1037/dev0000330.
Cisek, P. & Kalaska, J. F. (2010). Neural mechanisms for interacting with a world full of action choices. *Annual Review of Neuroscience*, *33*, 269–298. https://doi.org/10.1146/annurev.neuro.051508.135409.

Clark, A. & Chalmers, D. (1998). The extended mind. *Analysis*, *58*(1), 7–19.

Clark, C., Pritchard, V. E. & Woodward, L. J. (2010). Preschool executive functioning abilities predict early mathematics achievement. *Developmental Psychology*, *46*(5), 1176–1191. https://doi.org/10.1037/a0019672.

Clarke, S. & Beck, J. (2021). The number sense represents (rational) numbers. *Behavioral and Brain Sciences*, *44*, e178. https://doi.org/10.1017/S0140525X21000571.

Cole, J. C. (2009). Creativity, freedom, and authority: A new perspective on the metaphysics of mathematics. *Australasian Journal of Philosophy*, *87*(4), 589–608. https://doi.org/10.1080/00048400802598629.

(2013). Towards an institutional account of the objectivity, necessity, and atemporality of mathematics. *Philosophia Mathematica*, *21*(1), 9–36. https://doi.org/10.1093/philmat/nks019.

(2015). Social construction, mathematics, and the collective imposition of function onto reality. *Erkenntnis*, *80*(6), 1101–1124. https://doi.org/10.1007/s10670–014-9708-8.

Colyvan, M. (2001). *The Indispensability of Mathematics*. Oxford University Press.

Conde-Valverde, M., Martínez, I., Quam, R. M., Rosa, M., Velez, A. D., Lorenzo, C., Jarabo, P., Bermúdez de Castro, J. M., Carbonell, E. & Arsuaga, J. L. (2021). Neanderthals and Homo sapiens had similar auditory and speech capacities. *Nature Ecology & Evolution*, *5*(5), Article 5. https://doi.org/10.1038/s41559–021-01391-6.

Confer, J. C., Easton, J. A., Fleischman, D. S., Goetz, C. D., Lewis, D. M. G., Perilloux, C. & Buss, D. M. (2010). Evolutionary psychology: Controversies, questions, prospects, and limitations. *American Psychologist*, *65*(2), 110–126. https://doi.org/10.1037/a0018413.

Cutini, S., Scatturin, P., Moro, S. B. & Zorzi, M. (2014). Are the neural correlates of subitizing and estimation dissociable? An FNIRS investigation. *Neuroimage*, *85*, 391–399.

Davidson, K., Eng, K. & Barner, D. (2012). Does learning to count involve a semantic induction? *Cognition*, *123*, 162–173.

Davis, H. & Pérusse, R. (1988). Numerical competence in animals: Definitional issues, current evidence, and a new research agenda. *Behavioral and Brain Sciences*, *11*(4), 561–579. https://doi.org/10.1017/S0140525X00053437.

Dawkins, R. (2016). *The Extended Selfish Gene*, 40th anniversary ed. Oxford University Press.

De Bruin, L., Newen, A. & Gallagher, S. (Eds.). (2018). *The Oxford Handbook of 4E Cognition*. Oxford University Press.

De Cruz, H. (2016). Numerical cognition and mathematical realism. *Philosopher's Imprint*, *16*, 1–13.

De Cruz, H. & De Smedt, J. (2010). The innateness hypothesis and mathematical concepts. *Topoi*, *29*(1), 3–13.

De Cruz, H., Neth, H. & Schlimm, D. (2010). The cognitive basis of arithmetic. In B. Löwe & T. Müller (Eds.), *PhiMSAMP. Philosophy of Mathematics: Sociological Aspects and Mathematical Practice* (pp. 59–106). College Publications.

Decock, L. (2008). The conceptual basis of numerical abilities: One-to-one correspondence versus the successor relation. *Philosophical Psychology, 21*(4), 459–473. https://doi.org/10.1080/09515080802285255.

Dedekind, R. (1888). *Was sind und was sollen die Zahlen?: Stetigkeit und irrationale Zahlen* (S. Müller-Stach, Ed.; 1. Auflage). Springer Spektrum. https://doi.org/10.1007/978-3-662-54339-9.

Dehaene, S. (1997). *The Number Sense: How the Mind Creates Mathematics*, 1st ed. Oxford University Press.

(2001a). Précis of the number sense. *Mind & Language, 16*(1), 16–36.

(2001b). Subtracting pigeons: Logarithmic or linear? *Psychological Science, 12*(3), 244–246.

(2003). The neural basis of the Weber–Fechner law: A logarithmic mental number line. *Trends in Cognitive Sciences, 7*(4), 145–147.

(2005). Evolution of human cortical circuits for reading and arithmetic: The "neuronal recycling" hypothesis. In S. Dehaene, J. R. Duhamel, M. Hauser & G. Rizzolatti (Eds.), *From Monkey Brain to Human Brain* (pp. 133–157). MIT Press.

(2007). Symbols and quantities in parietal cortex: Elements of a mathematical theory of number representation and manipulation. *Sensorimotor Foundations of Higher Cognition, 22*, 527–574.

(2009). *Reading in the Brain: The New Science of How We Read*. Penguin.

(2011). *The Number Sense: How the Mind Creates Mathematics*, revised and updated ed. Oxford University Press.

(2020). *How We Learn: Why Brains Learn Better Than Any Machine ... for Now*. Viking.

Dehaene, S. & Akhavein, R. (1995). Attention, automaticity, and levels of representation in number processing. *Journal of Experimental Psychology: Learning, Memory, and Cognition, 21*(2), 314.

Dehaene, S., Bossini, S. & Giraux, P. (1993). The mental representation of parity and number magnitude. *Journal of Experimental Psychology: General, 122*(3), 371–396. https://doi.org/10.1037/0096-3445.122.3.371.

Dehaene, S. & Changeux, J. P. (1993). Development of elementary numerical abilities: A neuronal model. *Journal of Cognitive Neuroscience, 5*(4), 390–407.

Dehaene, S. & Cohen, L. (1995). Towards an anatomical and functional model of number processing. *Mathematical Cognition, 1*(1), 83–120.

Dehaene, S., Izard, V., Spelke, E. & Pica, P. (2008). Log or linear? Distinct intuitions of the number scale in Western and Amazonian indigene cultures. *Science, 320*, 1217–1220.

Dehaene-Lambertz, G. & Spelke, E. S. (2015). The infancy of the human brain. *Neuron, 88*(1), 93–109. https://doi.org/10.1016/j.neuron.2015.09.026.

Detlefsen, M., Erlandson, D. K., Heston, J. C. & Young, C. M. (1976). Computation with Roman numerals. *Archive for History of Exact Sciences*, *15*(2), 141–148.

Dos Santos, C. F. (2021). Enculturation and the historical origins of number words and concepts. *Synthese*, *199*, 9257–9287. https://doi.org/10.1007/s11229–021-03202-8.

(2022). Re-establishing the distinction between numerosity, numerousness, and number in numerical cognition. *Philosophical Psychology*, *35*(8), 1152–1180. https://doi.org/10.1080/09515089.2022.2029387.

Dostoevsky, F. (1864). *Notes from Underground* (R. Pevear & L. Volokhonsky, trans.). Vintage Classics.

Dummett, M. (1959). Wittgenstein's philosophy of mathematics. *The Philosophical Review*, *68*(3), 324–348. https://doi.org/10.2307/2182566.

(1978). *Truth and Other Enigmas*. Harvard University Press.

(2006). *Thought and Reality*. Oxford University Press.

Dutilh Novaes, C. (2012). *Formal Languages in Logic: A Philosophical and Cognitive Analysis*. Cambridge University Press.

Dutilh Novaes, C. & dos Santos, C. F. (2021). Numerosities are not ersatz numbers. *Behavioral and Brain Sciences*, *44*, e198. https://doi.org/10.1017/S0140525X21000984.

Enderton, H. B. (1977). *Elements of Set Theory*. Academic Press.

Epps, P. (2006). Growing a numeral system: The historical development of numerals in an Amazonian language family. *Diachronica*, *23*(2), 259–288. https://doi.org/10.1075/dia.23.2.03epp.

Euclid. (1956). *The Thirteen Books of Euclid's Elements. Vol. 1: Introduction and Books I, II*, 2nd ed. revised with additions, Vol. 1. Dover Publications.

Everett, C. (2017). *Numbers and the Making of Us: Counting and the Course of Human Cultures*. Harvard University Press.

Everett, C. & Madora, K. (2012). Quantity recognition among speakers of an anumeric language. *Cognitive Science*, *36*(1), 130–141. https://doi.org/10.1111/j.1551-6709.2011.01209.x.

Fabry, R. E. (2017). Cognitive innovation, cumulative cultural evolution, and enculturation. *Journal of Cognition and Culture*, *17*(5), 375–395.

(2018). Betwixt and between: The enculturated predictive processing approach to cognition. *Synthese*, *195*(6), 2483–2518.

(2020). The cerebral, extra-cerebral bodily, and socio-cultural dimensions of enculturated arithmetical cognition. *Synthese*, *197*, 3685–3720.

Fabry, R. E. & Pantsar, M. (2021). A fresh look at research strategies in computational cognitive science: The case of enculturated mathematical problem solving. *Synthese*, *198*(4), 3221–3263. https://doi.org/10.1007/s11229–019-02276-9.

Feferman, S. (2009). Conceptions of the continuum. *Intellectica*, *51*(1), 169–189.

Feigenson, L., Dehaene, S. & Spelke, E. (2004). Core systems of number. *Trends in Cognitive Sciences*, *8*(7), 307–314.

Fechner, G. T. (1860). Elements of psychophysics. In W. Dennis (Ed.), *Readings in the History of Psychology* (pp. 206–213). Appleton-Century-Crofts.
Ferreirós, J. (2016). *Mathematical Knowledge and the Interplay of Practices.* Princeton.
Field, H. (1980). *Science without Numbers.* Oxford University Press.
(1989). *Realism, Mathematics, and Modality.* Blackwell.
(2022). Conventionalism about mathematics and logic. *Noûs, 57*(4), 815–831. https://doi.org/10.1111/nous.12428.
FitzSimons, G. E. & Godden, G. L. (2000). Review of research on adults learning mathematics. In D. Coben, J. O'Donoghue & G. E. Fitzsimons (Eds.), *Perspectives on Adults Learning Mathematics*, Mathematics Education Library, Vol. 21 (pp. 13–45). Springer.
Fodor, J. (1980). Fixation of belief and concept acquisition. In M. Piattelli-Palmarini (Ed.), *Language and Learning: The Debate between Jean Piaget and Noam Chomsky* (pp. 142–149). Harvard University Press.
(1983). *The Modularity of Mind: An Essay on Faculty Psychology.* MIT Press.
(2010). Woof, woof. *Times Literary Supplement*, October 8, 7–8.
Frank, M. C., Everett, D., Fedorenko, E. & Gibson, E. (2008). Number as a cognitive technology. *Cognition, 108*, 819–824.
Frankopan, P. (2016). *The Silk Roads: A New History of the World.* Bloomsbury.
Freed, W. J., de Medinaceli, L., & Wyatt, R. J. (1985). Promoting functional plasticity in the damaged nervous system. *Science, 227*, 1544–1553.
Frege, G. (1879). Begriffsschift. In J. Heijenoort (Ed.), *From Frege to Gödel: A Source Book in Mathematical Logic, 1879–1931* (pp. 1–82). Harvard University Press
(1884). *Foundations of Arithmetic* (J. L. Austin, trans.). Blackwell.
(1892). Über Begriff und Gegenstand. *Vierteljahresschrift Für Wissenschaftliche Philosophie, 16*, 192–205.
Fuson, K. C. (1987). *Children's Counting and Concepts of Number.* Springer.
Fuson, K. C. & Secada, W. G. (1986). Teaching children to add by counting-on with one-handed finger patterns. *Cognition and Instruction, 3*(3), 229–260.
Gallagher, S. (2017). *Enactivist Interventions: Rethinking the Mind.* Oxford University Press.
Gallistel, C. R. (2017). Numbers and brains. *Learning & Behaviour, 45*(4), 327–328.
(2018). Finding numbers in the brain. *Philosophical Transactions of the Royal Society B: Biological Sciences, 373*(1740), 20170119. https://doi.org/10.1098/rstb.2017.0119.
Gallistel, C. R. & Gelman, R. (1992). Preverbal and verbal counting and computation. *Cognition, 44*(1–2), 43–74. https://doi.org/10.1016/0010-0277(92)90050-r.
Gallistel, C. R., Gelman, R. & Cordes, S. (2006). The cultural and evolutionary history of the real numbers. In S. Levinson & P. Jaisson (Eds.), *Evolution and Culture: A Fyssen Foundation Symposium* (pp. 247–274). MIT Press.

Garland, A. & Low, J. (2014). Addition and subtraction in wild New Zealand robins. *Behavioural Processes, 109*, 103–110. https://doi.org/10.1016/j.beproc.2014.08.022.

Geary, D. (2011). Cognitive predictors of achievement growth in mathematics: A five year longitudinal study. *Developmental Psychology, 47*(6), 1539–1552. https://doi.org/10.1037/a0025510.

Geary, D., Berch, D. & Mann Koepke, K. (Eds.). (2014). *Evolutionary Origins and Early Development of Number Processing*. Elsevier.

Gebuis, T., Cohen Kadosh, R. & Gevers, W. (2016). Sensory-integration system rather than approximate number system underlies numerosity processing: A critical review. *Acta Psychologica, 171*, 17–35. https://doi.org/10.1016/j.actpsy.2016.09.003.

Gelman, R. & Butterworth, B. (2005). Number and language: How are they related? *Trends in Cognitive Sciences, 9*(1), 6–10.

Gelman, R. & Gallistel, C. (1978). *The Child's Understanding of Number*. Harvard University Press.

(2004). Language and the origin of numerical concepts. *Science, 306*, 441–443.

Gibson, J. J. (1979). *The Ecological Approach to Visual Perception: Classic Edition*. Psychology Press.

Gödel, K. (1931). On formally undecidable propositions. In S. Feferman, J. Dawson, S. Kleene, G. Moore, R. Solovay & J. van Heijenoort (Eds.), *Collected Works: Vol. I: Publications 1929–1936* (pp. 145–195). Oxford University Press.

(1983). What is Cantor's continuum problem. In P. Benacerraf & H. Putnam (Eds.), *Philosophy of Mathematics* (pp. 470–485). Prentice-Hall.

Goldman, A. I. (1967). A causal theory of knowing. *Journal of Philosophy, 64*(12), 357–372. https://doi.org/10.2307/2024268.

Goodman, N. (1955). *Fact, Fiction, and Forecast*, 2nd ed. Harvard University Press.

Gordon, P. (2004). Numerical cognition without words: Evidence from Amazonia. *Science, 306*(5695), 496–499.

Gould, S. J. & Vrba, E. S. (1982). Exaptation: A missing term in the science of form. *Paleobiology, 8*(1), 4–15. https://doi.org/10.1017/S0094837300004310.

Greenfieldboyce, N. (2018). Math bee: Honeybees seem to understand the notion of zero. *NPR*. www.npr.org/2018/06/07/617863467/math-bee-honeybees-seem-to-understand-the-notion-of-zero.

Griffiths, P. E. (2001). What is innateness? *The Monist, 85*(1), 70–85. https://doi.org/10.5840/monist20028518.

Hadamard, J. (1954). *The Psychology of Invention in the Mathematical Field*. Dover Publications.

Halberda, J., Ly, R., Wilmer, J. B., Naiman, D. Q. & Germine, L. (2012). Number sense across the lifespan as revealed by a massive Internet-based sample. *Proceedings of the National Academy of Sciences, 109*(28), 11116–11120. https://doi.org/10.1073/pnas.1200196109.

Halberda, J., Mazzocco, M. M. M. & Feigenson, L. (2008). Individual differences in non-verbal number acuity correlate with maths achievement. *Nature, 455* (7213), 665–668. https://doi.org/10.1038/nature07246.

Hale, B., & Wright, C. (2001). *Reasons Proper Study.* Clarendon Press.
(2009). The metaontology of abstraction. In D. Chalmers, D. Manley & R. Wasserman (Eds.), *Metametaphysics: New Essays on the Foundations of Ontology* (pp. 178–212). Oxford University Press.
Hallett, M. (1988). *Cantorian Set Theory and Limitation of Size.* Clarendon Press.
Hauser, M. D., MacNeilage, P. & Ware, M. (1996). Numerical representations in primates. *Proceedings of the National Academy of Sciences, 93*(4), 1514–1517.
Heath, T. L. (1921). *A History of Greek Mathematics.* Dover Publications.
Hebbeler, J. (2015). Kant on necessity, insight, and a priori knowledge. *Archiv Für Geschichte Der Philosophie, 97*(1), 34–65. https://doi.org/10.1515/agph-2015-0002.
Heck, R. K. (2000). Cardinality, counting, and equinumerosity. *Notre Dame Journal of Formal Logic, 41,* 187–209.
Hellman, G. (1989). *Mathematics without Numbers.* Oxford University Press.
Hempel, C. G. (1945). On the nature of mathematical truth. *The American Mathematical Monthly, 52*(10), 543–556. https://doi.org/10.2307/2306103.
Henik, A. & Tzelgov, J. (1982). Is three greater than five: The relation between physical and semantic size in comparison tasks. *Memory & Cognition, 10*(4), 389–395. https://doi.org/10.3758/BF03202431.
Henrich, J. (2015). *The Secret of Our Success: How Culture is Driving Human Evolution, Domesticating our Species, and Making Us Smarter.* Princeton University Press.
Heyes, C. (2012). Grist and mills: On the cultural origins of cultural learning. *Philosophical Transactions of the Royal Society B: Biological Sciences, 367* (1599), 2181–2191. https://doi.org/10.1098/rstb.2012.0120.
(2018). *Cognitive Gadgets: The Cultural Evolution of Thinking.* Harvard University Press.
Hilbert, D. (1902). *The Foundations of Geometry.* Open Court.
Hinrichs, J. V., Yurko, D. S. & Hu, J. M. (1981). Two-digit number comparison: Use of place information. *Journal of Experimental Psychology: Human Perception and Performance, 7*(4), 890–901.
Hintikka, J. (1996). *Principles of Mathematics Revisited.* Cambridge University Press.
Hohol, M. (2019). *Foundations of Geometric Cognition.* Routledge.
Hohol, M., Wołoszyn, K., Nuerk, H.-C. & Cipora, K. (2018). A large-scale survey on finger counting routines, their temporal stability and flexibility in educated adults. *PeerJ, 6,* e5878. https://doi.org/10.7717/peerj.5878.
Howard, S. R., Avarguès-Weber, A., Garcia, J. E., Greentree, A. D. & Dyer, A. G. (2018). Numerical ordering of zero in honey bees. *Science, 360*(6393), 1124–1126. https://doi.org/10.1126/science.aar4975.
Hutchins, E. (1994). *Cognition in the Wild.* MIT Press.
Hutto, D. D. (2019). Re-doing the math: Making enactivism add up. *Philosophical Studies, 176,* 827–837.
Hutto, D. D. & Myin, E. (2013). *Radicalizing Enactivism. Basic Minds without Content.* MIT Press.

(2017). *Evolving Enactivism. Basic Minds Meet Content.* MIT Press.
Hyde, D. C. (2011). Two systems of non-symbolic numerical cognition. *Frontiers in Human Neuroscience, 5*, 150.
Hyde, D. C. & Ansari, D. (2018). Advances in understanding the development of the mathematical brain. *Developmental Cognitive Neuroscience, 30*, 236.
Ifrah, G. (1998). *The Universal History of Numbers: From Prehistory to the Invention of the Computer.* Harville Press.
Imbo, I., Duverne, S. & Lemaire, P. (2007). Working memory, strategy execution, and strategy selection in mental arithmetic. *Quarterly Journal of Experimental Psychology, 60*(9), 1246–1264. https://doi.org/10.1080/17470210600943419.
Irvine, A. D. (1989). *Physicalism in Mathematics.* Kluwer Academic Publishers.
Izard, V., Sann, C., Spelke, E. S. & Streri, A. (2009). Newborn infants perceive abstract numbers. *Proceedings of the National Academy of Sciences, 106*(25), 10382–10385. https://doi.org/10.1073/pnas.0812142106.
Izard, V., Streri, A. & Spelke, E. (2014). Toward exact number: Young children use one-to-one correspondence to measure set identity but not numerical equality. *Cognitive Psychology, 72*, 27–53.
Jones, M. (2020). Numerals and neural reuse. *Synthese, 197*, 3657–3681.
Kahneman, D., Treisman, A. & Gibbs, B. J. (1992). The reviewing of object files: Object-specific integration of information. *Cognitive Psychology, 24*(2), 175–219. https://doi.org/10.1016/0010-0285(92)90007-O.
Kallen, S. A. (2001). *The Mayans.* Lucent Books.
Kanjlia, S., Lane, C., Feigenson, L. & Bedny, M. (2016). Absence of visual experience modifies the neural basis of numerical thinking. *Proceedings of the National Academy of Sciences, 113*(40), 11172–11177. https://doi.org/10.1073/pnas.1524982113.
Kant, I. (1787). *Critique of Pure Reason.* Cambridge University Press.
Kaufman, E. L., Lord, M. W., Reese, T. W. & Volkmann, J. (1949). The discrimination of visual number. *The American Journal of Psychology, 62*, 498–525. https://doi.org/10.2307/1418556.
Kawai, M. (1965). Newly-acquired pre-cultural behavior of the natural troop of Japanese monkeys on Koshima islet. *Primates, 6*(1), 1–30. https://doi.org/10.1007/BF01794457.
Kitcher, P. (1983). *The Nature of Mathematical Knowledge.* Oxford University Press.
(1992). The naturalists return. *The Philosophical Review, 101*(1), 53–114. https://doi.org/10.2307/2185044.
Kline, M. (1973). *Why Johnny Can't Add: The Failure of the New Math.* St. Martin's Press.
Knops, A. (2020). *Numerical Cognition. The Basics.* Routledge.
Kövecses, Z. & Benczes, R. (2010). *Metaphor: A Practical Introduction*, 2nd ed. Oxford University Press.
Kripke, S. A. (1963). Semantical considerations on modal logic. *Acta Philosophica Fennica, 16*, 83–94.

(1980). *Naming and Necessity*. Blackwell Publishers.

(1982). *Wittgenstein on Rules and Private Language: An Elementary Exposition*. Harvard University Press.

Kuhn, T. S. (1993). Afterwords. In P. Horwich (Ed.), *Educational Theory* (pp. 311–341). MIT Press.

Kummer, H., Goodall, J. & Weiskrantz, L. (1985). Conditions of innovative behaviour in primates. *Philosophical Transactions of the Royal Society of London. B, Biological Sciences, 308*(1135), 203–214. https://doi.org/10.1098/rstb.1985.0020.

Lakoff, G. & Johnson, M. (2003). *Metaphors We Live by*. University of Chicago Press.

Lakoff, G. & Núñez, R. (2000). *Where Mathematics Comes from*. Basic Books.

Laland, K. N. (2017). *Darwin's Unfinished Symphony: How Culture Made the Human Mind*. Princeton University Press.

Landry, E. (2023). *Plato was Not a Mathematical Platonist. Elements in the Philosophy of Mathematics*. Cambridge University Press. https://doi.org/10.1017/9781009313797.

Lane, D. A. (2016). Innovation cascades: Artefacts, organization and attributions. *Philosophical Transactions of the Royal Society B: Biological Sciences, 371* (1690), 20150194. https://doi.org/10.1098/rstb.2015.0194.

Lange, M. (2017). *Because without Cause: Non-causal Explanations in Science and Mathematics*. Oxford University Press.

Lee, M. D. & Sarnecka, B. W. (2010). A model of knower-level behavior in number concept development. *Cognitive Science, 34*(1), 51–67.

(2011). Number-knower levels in young children: Insights from Bayesian modeling. *Cognition, 120*(3), 391–402.

Leibovich, T., Katzin, N., Harel, M. & Henik, A. (2017). From "sense of number" to "sense of magnitude": The role of continuous magnitudes in numerical cognition. *The Behavioral and Brain Sciences, 40*, e164. https://doi.org/10.1017/S0140525X16000960.

Leitgeb, H. (2020). Why pure mathematical truths are metaphysically necessary: A set-theoretic explanation. *Synthese, 197*(7), 3113–3120. https://doi.org/10.1007/s11229–018-1873-x.

Lemer, C., Dehaene, S., Spelke, E. & Cohen, L. (2003). Approximate quantities and exact number words: Dissociable systems. *Neuropsychologia, 41*(14), 1942–1958. https://doi.org/10.1016/S0028–3932(03)00123-4.

Leslie, A. M., Gelman, R. & Gallistel, C. R. (2008). The generative basis of natural number concepts. *Trends in Cognitive Sciences, 12*(6), 213–218. https://doi.org/10.1016/j.tics.2008.03.004.

Lewis, D. (1970). General semantics. *Synthese, 22*(1/2), 18–67.

Linnebo, Ø. (2018a). Platonism in the philosophy of mathematics. In E. Zalta (Ed.), *The Stanford Encyclopedia of Philosophy*. https://plato.stanford.edu/archives/spr2018/entries/platonism-mathematics.

(2018b). *Thin Objects*. Oxford University Press.

Lipton, J. S. & Spelke, E. S. (2003). Origins of number sense. Large-number discrimination in human infants. *Psychological Science, 14*(5), 396–401. https://doi.org/10.1111/1467-9280.01453.

Lumsden, C. & Wilson, E. (1981). *Genes, Mind, and Culture: The Coevolutionary Process*. Harvard University Press.

Maddy, P. (2014). A second philosophy of arithmetic. *The Review of Symbolic Logic*, *7*(2), 222–249.

Maguire, E. A., Gadian, D. G., Johnsrude, I. S., Good, C. D., Ashburner, J., Frackowiak, R. S. J. & Frith, C. D. (2000). Navigation-related structural change in the hippocampi of taxi drivers. *Proceedings of the National Academy of Sciences*, *97*(8), 4398–4403. https://doi.org/10.1073/pnas.070039597.

Maguire, E. A., Spiers, H. J., Good, C. D., Hartley, T., Frackowiak, R. S. J. & Burgess, N. (2003). Navigation expertise and the human hippocampus: A structural brain imaging analysis. *Hippocampus*, *13*(2), 250–259. https://doi.org/10.1002/hipo.10087.

Malafouris, L. (2013). *How Things Shape the Mind: A Theory of Material Engagement*. MIT Press.

Manders, K. (2008). The Euclidean diagram. In P. Mancosu (Ed.), *The Philosophy of Mathematical Practice* (pp. 80–133). Oxford University Press.

Margolis, E. & Laurence, S. (2008). How to learn the natural numbers: Inductive inference and the acquisition of number concepts. *Cognition*, *106*, 924–939.

Mateos-Aparicio, P. & Rodríguez-Moreno, A. (2019). The impact of studying brain plasticity. *Frontiers in Cellular Neuroscience*, *13*, 66. www.frontiersin.org/articles/10.3389/fncel.2019.00066.

McCloskey, M. (1992). Cognitive mechanisms in numerical processing: Evidence from acquired dyscalculia. *Cognition*, *44*(1–2), 107–157. https://doi.org/10.1016/0010-0277(92)90052-j.

McCloskey, M. & Macaruso, P. (1995). Representing and using numerical information. *The American Psychologist*, *50*(5), 351–363. https://doi.org/10.1037//0003-066x.50.5.351.

McCrink, K. (2015). Intuitive nonsymbolic arithmetic. In D. Geary, D. Herch & K. Koepke (Eds.), *Evolutionary Origins and Early Development of Number Processing* (pp. 201–223). Elsevier Academic Press. https://doi.org/10.1016/B978–0-12-420133-0.00008-9.

McCrink, K. & Wynn, K. (2004). Large-number addition and subtraction by 9-month-old infants. *Psychological Science*, *15*(11), 776–781. https://doi.org/10.1111/j.0956-7976.2004.00755.x.

McGarrigle, J. & Donaldson, M. (1974). Conservation accidents. *Cognition*, *3*(4), 341–350. https://doi.org/10.1016/0010-0277(74)90003-1.

Meck, W. H. & Church, R. M. (1983). A mode control model of counting and timing processes. *Journal of Experimental Psychology: Animal Behavior Processes*, *9*(3), 320.

Menary, R. (2014). Neuronal recycling, neural plasticity and niche construction. *Mind and Language*, *29*(3), 286–303.

(2015). *Mathematical Cognition: A Case of Enculturation*. MIND Group.

Menary, R. & Gillett, A. (2022). The tools of enculturation. *Topics in Cognitive Science*, *14*(2), 363–387. https://doi.org/10.1111/tops.12604.

Merzbach, U. C. & Boyer, C. B. (2011). *A History of Mathematics*, 3rd ed. John Wiley.

Mesoudi, A. & Thornton, A. (2018). What is cumulative cultural evolution? *Proceedings of the Royal Society B: Biological Sciences*, *285*(1880), 20180712. https://doi.org/10.1098/rspb.2018.0712.
Metzinger, T. (2013). The myth of cognitive agency: Subpersonal thinking as a cyclically recurring loss of mental autonomy. *Frontiers in Psychology*, *4*, 931. www.frontiersin.org/article/10.3389/fpsyg.2013.00931.
Michaelson, E. & Reimer, M. (2022). Reference. In E. N. Zalta (Ed.), *The Stanford Encyclopedia of Philosophy* (Summer 2022). Metaphysics Research Lab, Stanford University. https://plato.stanford.edu/archives/sum2022/entries/reference/.
Mill, J. S. (1843). A system of logic. In J. M. Robson (Ed.), *Collected Works of John Stuart Mill: Vols. 7 & 8*. University of Toronto Press.
Miller, K. F., Smith, C. M., Zhu, J. & Zhang, H. (1995). Preschool origins of cross-national differences in mathematical competence: The role of number-naming systems. *Psychological Science*, *6*(1), 56–60.
Morgan, C. L. (1894). *An Introduction to Comparative Psychology*. Palala Press.
Müller-Hill, E. (2009). Formalizability and knowledge ascriptions in mathematical practice. *Philosophia Scientiae*, *13*(2), 21–43.
Muthukrishna, M. & Henrich, J. (2016). Innovation in the collective brain. *Philosophical Transactions of the Royal Society B: Biological Sciences*, *371* (1690), 20150192. https://doi.org/10.1098/rstb.2015.0192.
Needham, J. & Wang, L. (1995). *Science and Civilisation in China*. Cambridge University Press.
Nelson, E. (2020). What Frege asked Alex the parrot: Inferentialism, number concepts, and animal cognition. *Philosophical Psychology*, *33*(2), 206–227. https://doi.org/10.1080/09515089.2019.1688777.
Netz, R. (1999). *The Shaping of Deduction in Greek Mathematics*. Cambridge University Press.
(2004). Eudemus of Rhodes, Hippocrates of Chios and the earliest form of a Greek mathematical text. *Centaurus*, *46*(4), 243–286.
Nieder, A. (2006). Temporal and spatial enumeration processes in the primate parietal cortex. *Science*, *313*, 1431–1435.
(2016). The neuronal code for number. *Nature Reviews Neuroscience*, *17*(6), 366.
(2019). *A Brain for Numbers: The Biology of the Number Instinct*, illustrated ed. The MIT Press.
Nieder, A. & Dehaene, S. (2009). Representation of number in the brain. *Annual Review of Neuroscience*, *32*, 185–208.
Nieder, A. & Miller, E. K. (2003). Coding of cognitive magnitude: Compressed scaling of numerical information in the primate prefrontal cortex. *Neuron*, *37*(1), 149–157. https://doi.org/10.1016/s0896-6273(02)01144-3.
Nissen, H. J., Damerow, P. & Englund, R. K. (1994). *Archaic Bookkeeping: Early Writing and Techniques of Economic Administration in the Ancient Near East* (P. Larsen, Trans.; 1st ed). University of Chicago Press.
Noël, M.-P. (2005). Finger gnosia: A predictor of numerical abilities in children? *Child Neuropsychology*, *11*(5), 413–430. https://doi.org/10.1080/09297040590951550.

Noles, N. S., Scholl, B. J. & Mitroff, S. R. (2005). The persistence of object file representations. *Perception & Psychophysics*, *67*(2), 324–334. https://doi.org/10.3758/BF03206495.

Núñez, R. E. (2009). Numbers and arithmetic: Neither hardwired nor out there. *Biological Theory*, *4*(1), 68–83. https://doi.org/10.1162/biot.2009.4.1.68.

(2011). No innate number line in the human brain. *Journal of Cross-Cultural Psychology*, *42*(4), 651–668.

(2017). Is there really an evolved capacity for number? *Trends in Cognitive Science*, *21*, 409–424.

Núñez, R. E., d'Errico, F., Gray, R. D. & Bender, A. (2021). The perception of quantity ain't number: Missing the primacy of symbolic reference. *Behavioral and Brain Sciences*, *44*, e199. https://doi.org/10.1017/S0140525X21001023.

Obayashi, S., Suhara, T., Kawabe, K., Okauchi, T., Maeda, J., Akine, Y., Onoe, H. & Iriki, A. (2001). Functional brain mapping of monkey tool use. *NeuroImage*, *14*(4), 853–861. https://doi.org/10.1006/nimg.2001.0878.

Ojose, B. (2008). Applying Piaget's theory of cognitive development to mathematics instruction. *Mathematics Educator*, *18*(1), 26–30.

Orwell, G. (1961). *1984*. Signet Classic.

Overmann, K. A. (2016). The role of materiality in numerical cognition. *Quaternary International*, *405*, 42–51.

(2018). Constructing a concept of number. *Journal of Numerical Cognition*, *4*(2), 464–493.

(2021a). Finger-counting in the Upper Palaeolithic. *Rock Art Research*, *31*(1), 63–80.

(2021b). Updating the abstract–concrete distinction in Ancient Near Eastern numbers. *Cuneiform Digital Library Journal*, *2018*(1), 1–22. https://cdli.mpiwg-berlin.mpg.de/articles/cdlj/2018-1.

Pantsar, M. (2009). Truth, Proof and Gödelian Arguments: A Defence of Tarskian Truth in Mathematics. PhD Thesis. University of Helsinki.

(2014). An empirically feasible approach to the epistemology of arithmetic. *Synthese*, *191*(17), 4201–4229. https://doi.org/10.1007/s11229–014-0526-y.

(2015a). Assessing the empirical philosophy of mathematics. *Discipline filosofiche*, *XXV*, 111–130. https://doi.org/10.1400/236780.

(2015b). In search of aleph-null: How infinity can be created. *Synthese*, *192*(8), 2489–2511.

(2016a). Frege, dedekind, and the modern epistemology of arithmetic. *Acta Analytica*, *31*, 297–318. doi: 10.1007/s12136-015-0280-x.

(2016b). The modal status of contextually a priori arithmetical truths. In F. Boccuni & A. Sereni (Eds.), *Objectivity, Realism, and Proof* (pp. 67–79). Springer.

(2018a). Early numerical cognition and mathematical processes. *THEORIA. Revista de Teoría, Historia y Fundamentos de La Ciencia*, *33*(2), 285–304.

(2018b). Mathematical explanations and mathematical applications. In *Handbook of the Mathematics of the Arts and Sciences* (pp. 1–16). Springer.

(2019a). Cognitive and computational complexity: Considerations from mathematical problem solving. *Erkenntnis*, *86*, 961–997.
(2019b). The enculturated move from proto-arithmetic to arithmetic. *Frontiers in Psychology*, *10*, 1454.
(2020). Mathematical cognition and enculturation: Introduction to the Synthese special issue. *Synthese*, *197*, 3647–3655. https://doi.org/10.1007/s11229-019-.
(2021a). Bootstrapping of integer concepts: The stronger deviant-interpretation challenge. *Synthese*, *199*, 5791–5814. https://doi.org/10.1007/s11229-021-03046-2.
(2021b). Descriptive complexity, computational tractability, and the logical and cognitive foundations of mathematics. *Minds and Machines*, *31*(1), 75–98.
(2021c). Objectivity in mathematics, without mathematical objects. *Philosophia Mathematica*, *29*(3), 318–352. https://doi.org/10.1093/philmat/nkab010.
(2021d). On the development of geometric cognition: Beyond nature vs. nurture. *Philosophical Psychology*, *35*, 595–616. https://doi.org/10.1080/09515089.2021.2014441.
(2023a). Developing artificial human-like arithmetical intelligence (and why). *Minds and Machines*, *33*, 379–396. https://doi.org/10.1007/s11023-023-09636-y.
(2023b). From maximal intersubjectivity to objectivity: An argument from the development of arithmetical cognition. *Topoi*, *42*(1), 271–281. https://doi.org/10.1007/s11245-022-09842-w.
(2023c). On radical enactivist accounts of arithmetical cognition. *Ergo*, *9*, 57. https://doi.org/doi.org/10.3998/ergo.3120.
(2023d). On what ground do thin objects exist? In search of the cognitive foundation of number concepts. *Theoria*, *89*(3), 298–313. https://doi.org/10.1111/theo.12366.
Parker, F. W. (1879). *Quincy Course in Arithmetic*. Andesite Press.
Peano, G. (1889). The principles of arithmetic, presented by a new method. In H. Kennedy (Ed.), *Selected Works of Giuseppe Peano* (pp. 101–134). University of Toronto Press.
Pelland, J.-C. (2018). Which came first, the number or the numeral? In S. Bangu (Ed.), *Naturalizing Logico-Mathematical Knowledge* (pp. 179–194). Routledge.
(2020). What's new: Innovation and enculturation of arithmetical practices. *Synthese*, *197*, 3797–3822. https://doi.org/10.1007/s11229-018-02060-1.
Penner-Wilger, M. & Anderson, M. L. (2013). The relation between finger gnosis and mathematical ability: Why redeployment of neural circuits best explains the finding. *Frontiers in Psychology*, *4*, 877. https://doi.org/10.3389/fpsyg.2013.00877.
Penner-Wilger, M., Waring, R. J. & Newton, A. T. (2014). Subitizing and finger gnosis predict calculation fluency in adults. *Proceedings of the Annual Meeting*

of the Cognitive Science Society, *36*(36), 1150–1155. https://escholarship.org/uc/item/4vv725r4.

Penrose, R. (1989). *The Emperor's New Mind: Concerning Computers, Minds and the Laws of Physics*. Oxford University Press.

(1994). *Shadows of the Mind: A Search for the Missing Science of Consciousness*. Oxford University Press.

Pepperberg, I. M. (2006). Grey parrot numerical competence: A review. *Animal Cognition*, *9*(4), 377–391. https://doi.org/10.1007/s10071–006-0034-7.

(2012). Further evidence for addition and numerical competence by a grey parrot (Psittacus erithacus). *Animal Cognition*, *15*(4), 711–717.

Pepperberg, I. M. & Gordon, J. D. (2005). Number comprehension by a grey parrot (Psittacus erithacus), including a zero-like concept. *Journal of Comparative Psychology*, *119*(2), 197–209. https://doi.org/10.1037/0735-7036.119.2.197.

Petersen, S. E. & Sporns, O. (2015). Brain networks and cognitive architectures. *Neuron*, *88*(1), 207–219. https://doi.org/10.1016/j.neuron.2015.09.027.

Piaget, J. (1965). *Child's Conception of Number*. Norton.

(1970). *Science of Education and the Psychology of the Child*. Viking Press.

Piazza, M. (2010). Neurocognitive start-up tools for symbolic number representations. *Trends in Cognitive Sciences*, *14*(12), 542–551. https://doi.org/10.1016/j.tics.2010.09.008.

Piazza, M., Facoetti, A., Trussardi, A. N., Berteletti, I., Conte, S., Lucangeli, D., Dehaene, S. & Zorzi, M. (2010). Developmental trajectory of number acuity reveals a severe impairment in developmental dyscalculia. *Cognition*, *116*(1), 33–41. https://doi.org/10.1016/j.cognition.2010.03.012.

Piazza, M., Pinel, P., Le Bihan, D. & Dehaene, S. (2007). A magnitude code common to numerosities and number symbols in human intraparietal cortex. *Neuron*, *53*, 293–305.

Pica, P., Lemer, C., Izard, V. & Dehaene, S. (2004). Exact and approximate arithmetic in an Amazonian indigene group. *Science*, *306*(5695), 499–503.

Pinel, P., Dehaene, S., Riviere, D. & LeBihan, D. (2001). Modulation of parietal activation by semantic distance in a number comparison task. *Neuroimage*, *14*(5), 1013–1026.

Plato. (1992). *The Republic* (G. M. A. Grube, trans.). Hackett Publishing Company.

Popper, K. R. (1972). *Objective Knowledge: An Evolutionary Approach*, revised ed. Oxford University Press.

Posth, C., Nakatsuka, N., Lazaridis, I., Skoglund, P., Mallick, S., Lamnidis, T. C., Rohland, N., Nägele, K., Adamski, N., Bertolini, E., Broomandkhoshbacht, N., Cooper, A., Culleton, B. J., Ferraz, T., Ferry, M., Furtwängler, A., Haak, W., Harkins, K., Harper, T. K., . . . Reich, D. (2018). Reconstructing the deep population history of Central and South America. *Cell*, *175*(5), 1185–1197.e22. https://doi.org/10.1016/j.cell.2018.10.027.

Power, T. G. (2013). *Play and Exploration in Children and Animals*. Psychology Press.

Prado, J., Mutreja, R., Zhang, H., Mehta, R., Desroches, A. S., Minas, J. E. & Booth, J. R. (2011). Distinct representations of subtraction and multiplication in the neural systems for numerosity and language. *Human Brain Mapping*, *32*(11), 1932–1947. https://doi.org/10.1002/hbm.21159.

Putnam, H. (1967). Mathematics without foundations. *Journal of Philosophy*, *64*(1), 5–22. https://doi.org/10.2307/2024603.

(Ed.). (1979). *Mathematics, Matter and Method*, 2nd ed. Cambridge University Press.

(Ed.). (1983). 'Two dogmas' revisited. In *Philosophical Papers: Volume 3: Realism and Reason* (pp. 87–97). Cambridge University Press. https://doi.org/10.1017/CBO9780511625275.007.

Quine, W. V. (1951). Two dogmas of empiricism. *Philosophical Review*, *60*(1), 20–43. https://doi.org/10.2307/2266637.

(1966). The scope of language of science. In *The Ways of Paradox and Other Essays* (pp. 215–232). Random House.

Quinon, P. (2021). Cognitive structuralism: Explaining the regularity of the natural numbers progression. *Review of Philosophy and Psychology*, *13*, 127–149. https://link.springer.com/article/10.1007/s13164-021-00524-x.

Rafi, Z. & Greenland, S. (2020). Semantic and cognitive tools to aid statistical science: Replace confidence and significance by compatibility and surprise. *BMC Medical Research Methodology*, *20*, 244. https://doi.org/10.1186/s12874-020-01105-9.

Rayo, A. (2013). *The Construction of Logical Space*. Oxford University Press.

(2015). Nominalism, trivialism, logicism. *Philosophia Mathematica*, *23*(1), 65–86.

Reck, E. & Schiemer, G. (2023). Structuralism in the philosophy of mathematics. In E. N. Zalta & U. Nodelman (Eds.), *The Stanford Encyclopedia of Philosophy*. Stanford University Press. https://plato.stanford.edu/archives/spr2023/entries/structuralism-mathematics/.

Reeve, R. & Humberstone, J. (2011). Five- to 7-year-olds' finger gnosia and calculation abilities. *Frontiers in Psychology*, *2*, 359. https://doi.org/10.3389/fpsyg.2011.00359.

Revkin, S. K., Piazza, M., Izard, V., Cohen, L. & Dehaene, S. (2008). Does subitizing reflect numerical estimation? *Psychological Science*, *19*(6), 607–614.

Rey, G. (2014). Innate and learned: Carey, mad dog nativism, and the poverty of stimuli and analogies (yet again). *Mind & Language*, *29*, 109–132.

Rips, L. J., Asmuth, J. & Bloomfield, A. (2006). Giving the boot to the bootstrap: How not to learn the natural numbers. *Cognition*, *101*(3), 51–60.

Rips, L. J., Bloomfield, A. & Asmuth, J. (2008). From numerical concepts to concepts of number. *Behavioral and Brain Sciences*, *31*(6), 623–642. https://doi.org/10.1017/S0140525X08005566.

Rosen, G. (2020). Abstract objects. In E. N. Zalta (Ed.), *The Stanford Encyclopedia of Philosophy*. Stanford University Press. https://plato.stanford.edu/archives/spr2020/entries/abstract-objects/.

Rugani, R., Fontanari, L., Simoni, E., Regolin, L. & Vallortigara, G. (2009). Arithmetic in newborn chicks. *Proceedings of the Royal Society B: Biological Sciences, 276*(1666), 2451–2460.

Rugani, R., Vallortigara, G., Priftis, K. & Regolin, L. (2015). Number-space mapping in the newborn chick resembles humans' mental number line. *Science, 347*(6221), 534–536. https://doi.org/10.1126/science.aaa1379.

Sarnecka, B. W. & Carey, S. (2008). How counting represents number: What children must learn and when they learn it. *Cognition, 108*(3), 662–674.

Sarnecka, B. W. & Gelman, S. (2004). Six does not just mean a lot: Preschoolers see number words as specific. *Cognition, 92*, 329–352.

Sarnecka, B. W. & Wright, C. (2013). The idea of an exact number: Children's understanding of cardinality and equinumerosity. *Cognitive Science, 37*(8), 1493–1506.

Sarrazin, J.-C., Giraudo, M.-D., Pailhous, J., Bootsma, R. J., & Giraudo, M.-D. (2004). Dynamics of balancing space and time in memory: Tau and kappa effects revisited. *Journal of Experimental Psychology. Human Perception and Performance, 30*(3), 411–430. https://doi.org/10.1037/0096-1523.30.3.411.

Saxe, G. B. (1982). Developing forms of arithmetical thought among the Oksapmin of Papua New Guinea. *Developmental Psychology, 18*(4), 583–594. https://doi.org/10.1037/0012-1649.18.4.583.

Schlaug, G., Jäncke, L., Huang, Y., Staiger, J. F. & Steinmetz, H. (1995). Increased corpus callosum size in musicians. *Neuropsychologia, 33*(8), 1047–1055. https://doi.org/10.1016/0028-3932(95)00045-5.

Schlimm, D. (2018). Numbers through numerals: The constitutive role of external representations. In S. Bangu (Ed.), *Naturalizing Logico-Mathematical Knowledge* (pp. 195–217). Routledge.

Schlimm, D. & Neth, H. (2008). Modeling ancient and modern arithmetic practices: Addition and multiplication with Arabic and Roman numerals. In B. C. Love, K. McRae & V. M. Sloutsky (Eds.), *Proceedings of the 30th Annual Conference of the Cognitive Science Society* (pp. 2097–2102). Cognitive Science Society.

Schmandt-Besserat, D. (1996). *How Writing Came About.* University of Texas Press.

Schneider, R. M., Brockbank, E., Feiman, R. & Barner, D. (2022). Counting and the ontogenetic origins of exact equality. *Cognition, 218*, 104952. https://doi.org/10.1016/j.cognition.2021.104952.

Schröder, E. (1873). *Lehrbuch der Arithmetik und Algebra für Lehrer und Studirende I.* Teubner.

Searle, J. R. (1997). *The Construction of Social Reality*, illustrated ed. Free Press.

(2010). *Making the Social World: The Structure of Human Civilization.* Oxford University Press.

Sfard, A. (2008). *Thinking as Communicating: Human Development, the Growth of Discourses, and Mathematizing.* Cambridge University Press.

Shapiro, L. & Spaulding, S. (2021). Embodied cognition. In E. N. Zalta (Ed.), *The Stanford Encyclopedia of Philosophy*. Stanford University Press. https://plato.stanford.edu/archives/fall2021/entries/embodied-cognition/.

Shapiro, S. (1996). Mathematical structuralism. *Philosophia Mathematica*, *4*(2), 81–82. https://doi.org/10.1093/philmat/4.2.81.

(1997). *Philosophy of Mathematics: Structure and Ontology*. Oxford University Press.

(2000). *Thinking about Mathematics*. Oxford University Press.

Shum, J., Hermes, D., Foster, B. L., Dastjerdi, M., Rangarajan, V., Winawer, J., Miller, K. J. & Parvizi, J. (2013). A brain area for visual numerals. *Journal of Neuroscience*, *33*(16), 6709–6715. https://doi.org/10.1523/JNEUROSCI.4558-12.2013.

Simon, T. J., Hespos, S. J. & Rochat, P. (1995). Do infants understand simple arithmetic? A replication of Wynn (1992). *Cognitive Development*, *10*, 253–269.

Skemp, R. R. (1987). *The Psychology of Learning Mathematics*, expanded American ed. Routledge.

Smart, J. J. C. (2017). The mind/brain identity theory. In E. N. Zalta (Ed.), *The Stanford Encyclopedia of Philosophy*. Stanford University Press. https://plato.stanford.edu/archives/spr2017/entries/mind-identity/.

Spelke, E. S. (2000). Core knowledge. *American Psychologist*, *55*(11), 1233–1243. https://doi.org/10.1037/0003-066X.55.11.1233.

(2011). Natural number and natural geometry. In S. Dehaene & E. Brannon (Eds.), *Space, Time and Number in the Brain* (pp. 287–318). Academic Press.

Spelke, E. S. & Kinzler, K. D. (2007). Core knowledge. *Developmental Science*, *10*(1), 89–96. https://doi.org/10.1111/j.1467-7687.2007.00569.x.

Srihasam, K., Mandeville, J. B., Morocz, I. A., Sullivan, K. J. & Livingstone, M. S. (2012). Behavioral and anatomical consequences of early versus late symbol training in macaques. *Neuron*, *73*(3), 608–619. https://doi.org/10.1016/j.neuron.2011.12.022.

Stalnaker, R. C. (2003). Reference and necessity. In R. C. Stalnaker (Ed.), *Ways a World Might Be: Metaphysical and Anti-metaphysical Essays* (pp. 165–187). Oxford University Press. https://doi.org/10.1093/0199251487.003.0010.

Starkey, P. & Cooper, R. G. (1980). Perception of numbers by human infants. *Science*, *210*(4473), 1033–1035.

Sterelny, K. (2003). *Thought in a Hostile World: The Evolution of Human Cognition*. Blackwell.

(2012). *The Evolved Apprentice: How Evolution Made Humans Unique*. MIT Press.

Stevens, S. S. (1939). On the problem of scales for the measurement of psychological magnitudes. *Journal of Unified Science (Erkenntnis)*, *9*, 94–99.

Stewart, I. (2006). *Letters to a Young Mathematician*. Basic Books.

Stjernfelt, F. & Pantsar, M. (2023). Peirce's philosophy of notations and the trade-offs in comparing numeral symbol systems. *Cognitive Semiotics*. https://doi.org/10.1515/cogsem-2023-2007.

Stoianov, I. & Zorzi, M. (2012). Emergence of a 'visual number sense' in hierarchical generative models. *Nature Neuroscience*, *15*(2), 194–196.

Stoljar, D. (2022). Physicalism. In E. N. Zalta (Ed.), *The Stanford Encyclopedia of Philosophy*. Stanford University Press. https://plato.stanford.edu/archives/sum2022/entries/physicalism/.

Strawson, P. F. (1992). *Analysis and Metaphysics: An Introduction to Philosophy*. Oxford University Press.

Stroud, B. (1965). Wittgenstein and logical necessity. *The Philosophical Review*, *74*(4), 504–518. https://doi.org/10.2307/2183126.

Sur, M., Garraghty, P. E. & Roe, A. W. (1988). Experimentally induced visual projections into auditory thalamus and cortex. *Science*, *242*(4884), 1437–1441. https://doi.org/10.1126/science.2462279.

Syrett, K., Musolino, J. & Gelman, R. (2012). How can syntax support number word acquisition? *Language Learning and Development*, *8*(2), 146–176. https://doi.org/10.1080/15475441.2011.583900.

Tait, W. W. (2001). Beyond the axioms: The question of objectivity in mathematics. *Philosophia Mathematica*, *9*(1), 21–36.

Tang, Y., Zhang, W., Chen, K., Feng, S., Ji, Y., Shen, J. & Liu, Y. (2006). Arithmetic processing in the brain shaped by cultures. *Proceedings of the National Academy of Sciences*, *103*(28), 10775–10780.

Tedlock, B. (1992). *Time and the Highland Maya*. University of New Mexico Press.

Testolin, A., Zou, W. Y. & McClelland, J. L. (2020). Numerosity discrimination in deep neural networks: Initial competence, developmental refinement and experience statistics. *Developmental Science*, *23*(5), e12940.

de Toffoli, S. (2023). Who's afraid of mathematical diagrams? *Philosopher's Imprint*, *23*, 9. https://doi.org/10.3998/phimp.1348.

Tomasello, M. (1999). *The Cultural Origins of Human Cognition*. Harvard University Press.

Tomasello, M., Kruger, A. C. & Ratner, H. H. (1993). Cultural learning. *Behavioral and Brain Sciences*, *16*(3), 495–552. https://doi.org/10.1017/S0140525X0003123X.

Trick, L. & Pylyshyn, Z. W. (1994). Why are small and large numbers enumerated differently? A limited capacity preattentive stage in vision. *Psychological Review*, *101*, 80–102.

Trivett, J. (1980). The multiplication table: To be memorized or mastered? *For the Learning of Mathematics*, *1*(1), 21–25.

Uller, C., Carey, S., Huntley-Fenner, G. & Klatt, L. (1999). What representations might underlie infant numerical knowledge? *Cognitive Development*, *14*(1), 1–36.

Valério, M. & Ferrara, S. (2022). Numeracy at the dawn of writing: Mesopotamia and beyond. *Historia Mathematica*, *59*, 35–53. https://doi.org/10.1016/j.hm.2020.08.002.

Vandervert, L. R. (2021). A brain for numbers: The biology of the number instinct by Andreas Nieder. *The Mathematical Intelligencer*, *43*(1), 123–127. https://doi.org/10.1007/s00283-020-10017-x.

Varela, F. J., Thompson, E. & Rosch, E. (1991). *The Embodied Mind: Cognitive Science and Human Experience*. MIT Press.

(2017). *The Embodied Mind: Cognitive Science and Human Experience*, revised ed. MIT Press.

Varga, S. (2019). *Scaffolded Minds: Integration and Disintegration*. MIT Press.

Varshney, L. R., Chen, B. L., Paniagua, E., Hall, D. H. & Chklovskii, D. B. (2011). Structural properties of the Caenorhabditis elegans neuronal network. *PLoS Computational Biology*, *7*(2), e1001066. https://doi.org/10.1371/journal.pcbi.1001066.

Verdine, B. N., Golinkoff, R. M., Hirsh-Pasek, K., Newcombe, N. S., Filipowicz, A. T. & Chang, A. (2014). Deconstructing building blocks: Preschoolers' spatial assembly performance relates to early mathematics skills. *Child Development*, *85*(3), 1062–1076. https://doi.org/10.1111/cdev.12165.

de Waal, F. B. M. (2017). *Are We Smart Enough to Know How Smart Animals Are?* Norton.

Warren, J. (2020). *Shadows of Syntax: Revitalizing Logical and Mathematical Conventionalism*. Oxford University Press.

Warren, M. (2018). Bees understand the concept of zero. *Science*. www.science.org/content/article/bees-understand-concept-zero.

Wasner, M., Nuerk, H.-C., Martignon, L., Roesch, S. & Moeller, K. (2016). Finger gnosis predicts a unique but small part of variance in initial arithmetic performance. *Journal of Experimental Child Psychology*, *146*, 1–16.

Way, E. C. (1991). *Knowledge Representation and Metaphor*. Kluwer Academic Publishers.

Whitehead, A. N. & Russell, B. (1910). *Principia Mathematica: Volumes 1–3*. Cambridge University Press.

Wiese, H. (2007). The co-evolution of number concepts and counting words. *Lingua*, *117*, 758–772.

Wilder, J. D., Kowler, E., Schnitzer, B. S., Gersch, T. M. & Dosher, B. A. (2009). Attention during active visual tasks: Counting, pointing, or simply looking. *Vision Research*, *49*(9), 1017–1031. https://doi.org/10.1016/j.visres.2008.04.032.

Wilson, J. & Clarke, D. (2004). Towards the modelling of mathematical metacognition. *Mathematics Education Research Journal*, *16*(2), 25–48. https://doi.org/10.1007/BF03217394.

Wittgenstein, L. (1972). *Philosophical Investigations* (G. E. M. Anscombe, Trans.; 3rd ed). Prentice Hall.

(1974). *Philosophical Grammar: Part I, The Proposition, and Its Sense, Part II, on Logic and Mathematics*. University of California Press.

(1976). *Lectures on the Foundations of Mathematics, Cambridge 1939* (C. Diamond, Ed.). University of Chicago Press.

(1978). *Remarks on the Foundations of Mathematics* (G. H. von Wright, Trans.; revised ed). MIT Press.

Wright, C. (1992). *Truth and Objectivity*. Harvard University Press.

Wynn, K. (1990). Children's understanding of counting. *Cognition*, *36*(2), 155–193.

(1992). Addition and subtraction by human infants. *Nature*, *358*, 749–751.

Xu, F. & Spelke, E. S. (2000). Large number discrimination in 6-month-old infants. *Cognition*, *74*(1), B1–B11. https://doi.org/10.1016/S0010–0277(99)00066-9.

Xuan, B., Zhang, D., He, S. & Chen, X. (2007). Larger stimuli are judged to last longer. *Journal of Vision*, *7*(10), 2. https://doi.org/10.1167/7.10.2.

Yablo, S. (2002). Abstract objects: A case study. In A. Bottani, M. Carrara & P. Giaretta (Eds.), *Individuals, Essence and Identity: Themes of Analytic Metaphysics* (pp. 163–188). Springer Netherlands. https://doi.org/10.1007/978-94-017-1866-0_7.

Zach, R. (2019). Hilbert's program. In E. N. Zalta (Ed.), *The Stanford Encyclopedia of Philosophy*. Stanford University Press. https://plato.stanford.edu/entries/hilbert-program/.

Zahidi, K. (2021). Radicalizing numerical cognition. *Synthese*, *198*(Suppl 1), 529–545.

Zahidi, K. & Myin, E. (2016). Radically enactive numerical cognition. In G. Etzelmüller & C. Christian (Eds.), *Embodiment in Evolution and Culture* (pp. 57–72). Mohr Siebeck.

(2018). Making sense of numbers without a number sense. In S. Bangu (Ed.), *Naturalizing Logico Mathematical Knowledge* (pp. 218–233). Routledge.

Zebian, S. (2005). Linkages between number concepts, spatial thinking, and directionality of writing: The SNARC effect and the REVERSE SNARC effect in English and Arabic monoliterates, biliterates, and illiterate Arabic speakers. *Journal of Cognition and Culture*, *5*, 165–190. https://doi.org/10.1163/1568537054068660.

Zhang, J., & Norman, D. A. (1994). Representations in distributed cognitive tasks. *Cognitive Science*, *18*(1), 87–122. https://doi.org/10.1207/s15516709cog1801_3.

Index

Printed in the United States
by Baker & Taylor Publisher Services